WITNESSES
FOR
OIL

WITNESSES FOR OIL

The case against dismemberment

Congressional testimony and papers by R. J. Boushka; DeWitt Buchanan; Michael E. Canes; Annon M. Card; C. Howard Hardesty, Jr.; John E. Kasch; Donald C. O'Hara; Carel Otte; Walter R. Peirson; L. C. Soileau III; W. T. Slick, Jr.; The Southern Caucus; Standard Oil Company of California; Fred F. Steingraber; W.P. Tavoulareas; Charles J. Waidelich; Robert J. Welsh, Jr.; Hastings Wyman, Jr.

Compiled by
Michael E. Canes

Patricia Maloney Markun
Editor

James E. Coble
Consulting Editor

American Petroleum Institute

Printed in the United States of America

FIRST PRINTING

International Standard Book Number 0-89364-001
Library of Congress Catalog Card Number 76-26729

For information address:

American Petroleum Institute
2101 L Street, Northwest
Washington, D.C. 20037

Contents

List of Exhibits, Tables and Graphs

Introduction

Since late 1973, OPEC nations have skyrocketed the price of the oil we import. And the Arab oil embargo of 1973-1974 brought about the first peacetime shortfall of gasoline supplies in this country.

Some members of the Congress of the United States sought to put the blame for the long gasoline station lines and the fourfold increase in foreign oil prices on American oil companies. Their proposed legislation would dismember this country's oil companies in either of two ways. Some proposals would divide the large integrated companies into separate, single-function units. Other proposals would prevent the oil companies—our nation's most experienced energy producers—from participating in the development of other energy resources.

Hearings on these bills by both Houses of Congress have brought vigorous dissent from the executives of the petroleum companies—both large and small. Their testimony has been brought together in this volume. Some exhibits have been updated but, with minor exceptions, the wording has not been changed.

Selections by attorney-writer Hastings Wyman, Jr. on the Standard Oil breakup of 1911 and economist Michael E. Canes, both of the American Petroleum Institute, were not presented as testimony before the Congress. They are included to round out historically and technically the case against dismemberment.

This collection of papers offers careful analyses of all available data as evidence that the existing structure of the oil business is both competitive and efficient. Hopefully, this volume will help focus future debate on some of the more important reasons why the freedom to choose the most efficient and effective type of organization should be preserved, not only for oil enterprises but for all other American industrial enterprises as well.

F. N. Ikard, President
American Petroleum Institute

1

A View From the Independents

TESTIMONY BY RICHARD J. BOUSHKA, PRESIDENT, VICKERS ENERGY CORPORATION, BEFORE THE SUBCOMMITTEE ON ANTITRUST AND MONOPOLY, COMMITTEE ON THE JUDICIARY, UNITED STATES SENATE — JANUARY 27, 1976

Summary: There are no corporations able financially to buy the broken pieces in case forced divestiture comes to pass. Rather than dismantling the major oil companies by either vertical or horizontal divestiture, the government should encourage them to research all new domestic energy sources, since they are the ones with the applicable expertise. If they had been broken up years ago, we would not have had the phenomena of high-volume, low cost independent gasoline operations; thus prices to consumers would have been higher. Forced divestiture would eventually breed inefficient operators and high cost suppliers. The independents feel they can effectively compete with the majors in their present corporate structure.

Richard J. Boushka speaks . . .

Vickers Energy Corporation has 7,000 barrels per day of crude oil production and 100,000,000 cubic feet per day of natural gas production. We also own a 60,000 barrel per day refinery in Ardmore, Oklahoma, and market through 950 service stations, one-half of which are company-owned and operated. We are the 29th ranked oil company in the United States based on gasoline sales. These figures are significant but certainly minimal when compared to the vital statistics of the major oil companies. While at first blush one could immediately assume that we would profit by the breaking up of the majors, it is our strong opinion that not only is the converse true but we, as one of the intended beneficiaries, might suffer a fate worse than those who are forced to divest.

Before discussing the pros and cons of divestiture or divorcement, I should first like to discuss the practicality of the whole issue of vertical divestiture. If the majors are forced to sell pieces of their businesses, who can possibly afford to buy them? New refineries cost about $3,000 per daily barrel of capacity and used refineries approximately $2,000. Utilizing the lower figure for 500,000 barrels of capacity, one divested refinery would sell for $1,000,000,000. The refinery capacity proposed to be divested is considerably in excess of the 500,000 barrel illustration I used.

We and the other independents certainly could not step forward and pick up many of their divested assets. Most of the other large U.S. corporations have their own capital and financial problems at the moment. Even if the majors are willing or are forced to carry the financing, that loan would destroy most purchasers' balance sheet and curtail their other corporate activities for many years. I seriously question the practicality of having the majors sell their assets, and I am afraid that no one has really given this aspect serious consideration. The only alternative would be to spin the assets they are forced to divest to their existing stockholders.

Let us take a look at the refining and marketing company which has been set out in the cruel world by itself without preparation or warning. For years the majors have been accused of subsidizing their refining and marketing operations from their vast crude oil holdings and profits. I cannot verify the accuracy of that accusation, but I believe it to be true. If it is, the new divested refining and marketing company will require two things: one, new money poured into it to make it viable, and two, higher prices from the consumer to offset the historical subsidy. There is no question in my mind that standing alone, the refining and marketing divisions of the large oil companies will have to raise prices to receive an adequate return on stockholder investment and sufficient capital to pay dividends and maintain their existing properties. If the consumer must pay higher prices to keep the various pieces in existence, obviously you will have defeated one of your basic divestiture goals.

Rather than be distraught by the huge size and power of the major oil companies, it seems to me that the American public ought to be thankful they exist in their present form. There are some projects that independents just simply cannot handle. For example, just a few months ago when the final drilling was completed on the federal leases, offshore Florida, the final score unfortunately was no oil, no gas; and the companies, primarily majors, that participated in that exploration program lost a combined total of $1,000,000,000. You did not see Vickers on that list of losers. We too were impressed with the geologic possibilities that existed prior to drilling, but it was far too risky for us to expend a large portion of our net worth on one single play. The American public would have been the eventual benefactor if the operation had been successful but only the majors had the financial stamina for the actual results. They wrote it off and went about their business bruised but not destroyed. We need that kind of resiliency. If split up, will they retain that ability? It seems to us it is an unnecessary risk to take.

Although you are dealing with the subject of vertical divestiture in these hearings, the above point applies similarly to horizontal structure problems. If you break up the major oil companies and dilute their resources and expertise, the horizontal divestiture problem might become academic. Rather than restrict the majors from entering new energy forms like nuclear and coal, the government should be encouraging them to expend as many dollars as possible to speed the process of moving from dependency on foreign oil to new domestic energy sources, whatever they are. A fragmented approach to the development of new forms of energy will not only cost the American consumer more in the long run, it will also disrupt one of the nations most potent forces for seeking alternative energy sources. If a unified approach utilizing the tremendous resources of the major oil companies is not implemented and quickly, we as a nation may find ourselves in a period when we are phasing out of oil and gas with no ready alternative available. Especially at this time in our country's history, we cannot afford to dismantle its leading energy companies. Our business is our own, but other countries have actually encouraged mergers to insure that their industry would be in the hands of efficient, well-run, internationally competitive companies. It is ironic to me that we are thinking of going totally in the other direction.

If you agree with my position that we need the major oil company size and power for research and development of oil, gas, coal and nuclear energy, let me alleviate your concerns that if the majors are left alone they will eventually eliminate the independents in the oil industry.

There is a misconception regarding the power of the major oil companies in the marketplace. Specifically, I would like to address myself this morning to retail gasoline marketing where we have successfully competed with the majors for many years. When you drive through a town or city, more often than not the major oil prices will be $.02 to $.05 per gallon higher than the independent stations. There is a good sound basic reason for this with nothing sinister involved. In most cases, major oil company marketing costs are higher than the independents. I emphasize, in most cases, they are not the low cost operators. You might even say that some aspects of their system of marketing are antiquated. Because of these higher costs, the majors have lost market position over the past five to ten years. That is predictable for the high cost operator. It is apparent to me and hopefully to you that this is a situation that should be left alone. The competitive forces of the marketplace are working. Any disturbance of these natural forces might cause this trend to reverse and actually hurt the independent segment of the industry, the very thing a breakup is proposed to help. When the independent sector is hurt, the end-loser is the consumer, and even though he might not realize it, he has received a bargain for many years due to the very competitive nature of the gasoline marketing process.

What we are asking is that the natural forces of the market be left alone. Divestiture will not make overnight low cost operators of the majors.

In the same vein, if the large major oil companies had been broken up, say fifteen years ago, as is proposed today, the phenomenon of high-volume, low cost independent operators gaining market share would never have happened. Many of the large independents today exist because of the overbuilding of refining facilities by the majors, even though the overall costs of operation are not too much different in a 300,000 barrel per day major refinery than a 50,000 barrel per day independent facility. The incremental production from a large facility can be quite low cost. This supply of incremental or spot gasoline provided the springboard or basis for the construction of high volume, low cost units by many independents across the country. Two things would have happened if the majors had been broken up. First of all the over capacity might never have existed because instead of the one 300,000 barrel per day plant in my example, there might have been eight 30,000 barrel per day plants, and secondly, the surplus gasoline if it existed at all would not have represented the bargain it did to the independent and eventually the U.S. consumer. If the majors are broken up, we are actually encouraging the construction of smaller, less efficient refinery units. This will mean that there will be less gasoline available for the independent marketers who rely on others for economic sources of supply.

I noticed that one of the bills under consideration speaks exclusively to pipeline divestiture. Traditionally, the pipeline area causes the most emotional conversation and is the first discussion topic regarding divestiture. We feel the pipeline subject is overplayed, and a simple business made mysterious. Alternatives are too easy to come by for anyone to be injured over a long period of time.

I have heard some proposals which suggest that the majors divest themselves of their pipeline interests to independents or smaller companies. The one unanswered question that I have and I think you should have also, is who is going to take care of the majors' dealers and small business customers who have been supplied for years through the lines? If the new owners of the line curtail the shipment of the former owners, the majors would not be hurt proportionately nearly so badly as their independent dealers and jobbers. We feel the general pipeline situation has provided the logistics for effective competition, or companies such as ours would not have been able to grow as we have over the past five to ten years. Much like the service station area, any disruption could have the opposite effect than is intended by divestiture.

There may be those in the oil industry who would disagree with my testimony here this morning. I ask them to examine their corporate consciences and determine whether or not they might be guilty of trying to justify past poor management decisions by blaming the majors. Gentlemen, in business as in politics, all important decisions involve risk. The most obvious risks in business are economic, but there are others that an effective manager must consider when charting a direction for his company. When Vickers decided to expand its refinery in Ardmore, Oklahoma, and did not own or control the volume of crude oil necessary to supply that new capacity, we assumed the calculated risk that we would be able to obtain that raw material in future years.

If I come to Washington three years from now, it should not be my prerogative to scream anti-major slogans regarding power or monopoly and request relief in whatever form, because I knew full well when I expanded that new refinery the risks and dangers involved in crude supply. Too many times, gentlemen, you as legislators are asked to find a solution to someone's predicament which was caused by his poor management judgment and decisions. When you take action in this situation, you penalize those who exercised good thinking and prudent planning.

Inherent in a discussion concerning the breaking up of the majors is that others in the oil industry need a subsidy to exist. We maintain that the free market in the oil industry will allow large, small, and medium sized companies all to coexist without subsidies which breed weakness and mediocrity. We feel that the recipient of a subsidy eventually becomes so dependent on that handout that he becomes an inefficient operator and a high cost supplier of products to the consumer.

We have great respect for the overall posture and power of the major oil companies. At the same time, we feel we can effectively compete with them in their present corporate style and format. Individually, I am extremely concerned as a private citizen, and corporately, we feel very strongly that any move to cut the majors "down to size" would be counterproductive and extremely detrimental to the best interests of the American consumer.

2

A View From A Large Oil Company

TESTIMONY BY W. T. SLICK, JR., SENIOR VICE PRESIDENT, EXXON COMPANY, U.S.A., BEFORE THE SUBCOMMITTEE ON ANTITRUST AND MONOPOLY, COMMITTEE ON THE JUDICIARY, UNITED STATES SENATE — NOVEMBER 12, 1975.

Summary: Vertical integration is an efficient form of industrial organization commonly used by many companies. It allows oil companies which are highly capital intensive and conduct very complex operations to run their businesses more smoothly and efficiently by reducing uncertainty and providing reliability of supply, product quality, price, and service. Vertical integration allows smaller inventories and less working capital, lower intermediate transaction costs, lower capital costs, and thus lower prices to consumers. Both large and small firms are integrated, but none is fully integrated in all functions. Divestiture would remove vertical integration, thus raising costs to consumers. It would hinder capital investment in energy resources, thus leading to increased dependence on imports, which in turn would strengthen OPEC and endanger the savings of the investing public.

*W. T. Slick speaks** . . .

The structure, size, and operating characteristics of oil companies are natural economic developments which provide petroleum products for the consumer at the lowest cost possible. Dismemberment of the petroleum industry would have severe repercussions for the U.S. economy and consumers.

Vertical integration is an efficient form of industrial organization commonly used in major U.S. industries by many companies, both large and small, in order to serve the consumer at the lowest possible cost. Contractual arrangements will be presented as a costly substitute for the inter-functional planning and coordination of integrated firms.

*Exhibits referred to by each witness appear at the end of the chapter.

Petroleum companies are logical candidates for a vertically integrated structure because of the capital intensive nature of the business and the complexity of operations. Many firms of all sizes are integrated, not just large firms. Contrary to general belief, vertically integrated firms do not operate closed integrated systems. In fact, evidence is presented to show that oil and products enter and leave the system in huge quantities at all levels.

Intense competition exists at all horizontal levels of the industry. Concentration is low in absolute and relative terms in each of the functions of the industry. Ease of entry into the industry is exhibited through the growth pattern of independents in all segments. Joint operations and exchange agreements are factors in reducing consumer costs. Profits, the final measure of competition in an industry, have been historically no higher for the petroleum industry than for the average of all U.S. manufacturing. Finally, the profit trends of 1975 fully corroborate the extraordinary circumstances of the years 1973 and 1974.

Dismemberment would have a severe impact on the economy and on U.S. consumers. In the short run, reduced capital spending would increase unemployment and reduce real economic growth. In the longer range, less domestic energy supplies would be available to the nation and the severity of economic impact would increase. In addition, the efficiencies of vertical integration would be lost, leaving the consumer with higher prices and less assurance of supply and the investor with capital losses.

The claim that dismemberment of U.S. oil companies would somehow weaken OPEC is shown to be false; it actually would strengthen OPEC, and would increase the risk of confrontation.

The Efficiencies of Vertical Integration

Vertical integration is an efficient form of industrial organization which is commonly used by a great number of American companies, both large and small, in order to better serve the consumer at lower costs. In addition to petroleum, some other industries which have integrated companies include autos, food, newspapers, steel, tobacco, beer, machinery, communications, paper, nonferrous metals, chemicals, rubber products, textiles, and stone, glass, and clay.

For instance, many newspapers and publishers have integrated backward into forest product companies and paper manufacturing companies to assure an adequate supply of newsprint. Many newspapers also have company-owned transportation equipment to insure efficient distribution of their product. Moreover, integration also takes the form of syndication whereby certain papers put feature writers under contract and distribute the stories exclusively to syndicate members.

Just as in the petroleum industry and the newspaper industry, vertical integration allows the companies in many other industries to deal with uncertainty in a company's economic environment by providing for reliability of supply, product quality, price, and service. Vertical integration also allows better coordination of expectations and activities between stages of production with a high degree of interdependency and, therefore, results in reductions of required inventories and working capital. Moreover, these aspects of vertical integration improve a company's access to capital markets—not only in terms of a lower cost of capital because of lower risk, but also because of its increased ability to communicate to investors the entire interdependent investment picture.

Vertical integration also effects savings in transactions and communications costs in terms of reductions in operating and support staffs, and the concomitant elimination or reduction in reports and data handling, reduced costs of negotiations, reduced need for settlement of disputes, and better communication because of common training, experience, linguistic code, and community of interests among employees. All of these advantages of vertical integration obviously add up to cost savings to the consumer.

Efficient supply for a petroleum company, as for many companies in other integrated industries, requires large investments in special-purpose, long-life equipment. Thus, optimal investment considerations favor the award of long-term contracts to permit the supplier to amortize his investment with confidence.

Although long-term contracts can result in an industrial structure very similar to that existing under vertical integration, long-term contracts are not adequate substitutes for vertical integration. Contingent supply relations must be exhaustively stipulated in long-term contracts to prevent conflict over contract ambiguities. Unfortunately, this is often precluded by the bounds of rationality and the inability to foresee the future. Those stipulations which are specified are often costly in themselves. Suppliers demand a substantial premium when they commit their resources to a particular buyer under specified conditions for a long period of time. Moreover, costs multiply when changing technology and market circumstances force renegotiation. Needed amendments are costly—not only in terms of time and money, but also in the distortions in the decision-making processes they invariably create. Such distortions can include the total cancellation of plans because of the inability of a party to honor a long-term contract. Many industries have experienced severe hardships when suppliers have been unable to deliver vital raw materials.

The foregoing problems are taken from real world experience and are the factors which have motivated firms in many businesses to integrate either forward or backward and sometimes both.

Vertical integration is a logical and effective structure for the petroleum industry. The petroleum industry is characterized by large capital investments and complex operations, both physically and in the business sense. Vertical integration has been the logical and natural outgrowth of firms seeking to efficiently utilize capital intensive facilities to smoothly and effectively conduct their operations so they can deliver products to the consumer in a timely and economic manner.

These operations include a multitude of discrete physical and commercial transactions including gravity, magnetics, and seismic surveys; geological and geophysical studies; lease purchases; exploratory and development drilling; construction and installation of producing facilities; production, gathering, and transportation of crude and natural gas, and the construction of the required facilities; design, construction, and operation of gas processing facilities and refineries; storage of crude and products; sale of crude not compatible with processing facilities and the purchase of substitutes and supplemental volumes; design of product quality to customer specifications or requirements; product quality control; packaging and bulk transportation of products; distribution to retail sales points; and, finally, consumer sales with the associated billing and collection processes.

Each of these more than twenty-five discrete activities is in itself surrounded by significant planning and involves multiple decision-making steps. Vertical integration is merely the interrelating of these processes through a unified management organization and system. The efficiency that can result is obvious. The beneficiary is the consumer.

Each of these activities, being discrete, could be performed by a separate organization. In fact, many specialty firms do exist which limit their operations to but a few or even one of these steps. Many such firms have and will continue to prosper as long as they perform a useful economic function. But the size and complexity of the task at hand also requires large and complex integrated companies.

All elements of the petroleum industry—producing, refining, transportation, and marketing—are highly capital intensive. The cost of a new grass roots 250 MB/D refinery is in excess of $500 million, an installed offshore producing platform and facilities would range from $45 million to over $200 million, and the cost of a 20 percent interest in the Trans-Alaskan Pipeline would be in the neighborhood of $1.5 billion. In 1974, Exxon USA had $243,000 in fixed assets investment for each employee—an investment some seven times larger than the $33,000 median investment for all industries. It is therefore critical that large companies operate at a high level of capacity with minimum interruption in order to utilize their facilities efficiently and economically.

Historically domestic refiners have operated in the range of 85-93 percent utilization of capacity and have thereby struck a proper balance between the avoidance of high cost idle capacity on one hand and the danger of too little capacity on the other. The consumer benefits from the assurance of continued supply of refined products at the lowest possible production cost. The achievement of this type of pro-consumer performance is facilitated by the fact that the operators of over 96 percent of the U.S. refining capacity have removed at least some of the uncertainties surrounding raw material supply from the producing end, or the timely and effective product distribution from the marketing end by integrating either backward or forward. Vertical integration not only serves this function, but does it without introducing the vagaries and potential defaults inherent in third party contractual obligations. Consumers awaiting critical petroleum supplies to fuel industrial furnaces, hospitals, schools, and homes would be unsympathetic to the explanation that a third party entrepreneur was unable to fulfill his contractual obligation to produce the crude or transport the products.

The role of pipelines. Pipelines require large capital investments which can be minimized by the economies of scale. Pipelines require the commitment of sizable capital investments to long-life equipment, which is limited basically to one specialized use. If properly sized and constructed in a timely fashion, pipelines are usually the lowest cost transportation alternative to move crude to market. In most cases, crude pipelines would not be built without the financial support of the producing companies. When a new field is discovered there is much uncertainty as to ultimate reserve size and producing rate. Both of these factors are essential to determining both the required size and the probable economics of a pipeline. In such cases the producer has several choices. He can delay construction of transportation facilities until ultimate reserve size is established by shutting in production or, in the meantime, use an expensive alternative such as trucking. Alternatively, he can seek out an independent investor to build a pipeline. Such an investor would, however, need assurances of the size of the reserve to permit designing of a pipeline and assurances of throughput during the time required to recover his investment. The producer would have to assume not only the risks of the unknown dimensions of the discovery and its future producing history, but also the hazard risks of the pipeline operations of a third party. Given these alternatives, the producer as often as not would choose to construct the pipeline himself, thus having some check over controlling operating costs and avoiding underwriting the profits of another's investment.

In certain cases, such as TAPS, the investment in the pipeline can be extraordinarily high. Situations such as TAPS are further complicated by the problems of working in a new, relatively hostile environment in which it is difficult to forecast construction costs accurately at an early stage of project development. The increase in the expected cost of TAPS from initial estimates of $900 million to almost $7 billion is proof of the uncertainties associated with this type project.

It is not clear how the managements of producing companies and pipeline companies could structure throughput contracts that would provide the financial security demanded by backers, while at the same time maintaining stewardship of the financial obligations that producing management would owe to its stockholders. At best, development of such relationships would be a time consuming process that would delay bringing the crude to market. At worst, government intervention in terms of financial backing of the project could be required that would further add to the costs and complexities of the pipeline construction and operation.

Although the Alaskan North Slope discoveries are unique in many ways, they are believed to signal the trend of the future as most experts forecast that a large portion of the country's future crude reserve potential will be located in similar hostile environments, either in Alaska, or in deeper offshore waters. In either case, providing the transportation to move the crude to market will be a major factor in successful development of the crude potential. Producer participation is essential to the timely development of that transportation.

As suggested earlier, the complexity of modern petroleum operations requires a close and continuing functional interaction on scheduling and control. The potential value for this close coordination can best be appreciated by examining the operations of a large complex operation such as Exxon USA. Each day in 1974 Exxon produced on average 700,000 barrels per day of crude and condensate, purchased 113,000 barrels per day from royalty owners, purchased 737,000 barrels per day from other domestic producers, and imported 257,000 barrels per day from foreign sources; and, in turn, sold 780,000 barrels per day to other domestic users. Our domestic production occurs at over 10,000 individual wells scattered throughout 23 states in the U.S. Imports are received in the Gulf of Mexico and on the East and West Coasts.

After production, coordination must be arranged for the following materials, processes, products, and facilities:

Number of refineries/plants	7
Number of processing units	184
Number of different types of processes	49
Number of different crudes used	100
Grades of petroleum products manufactured	1,300
Bulk plants	921
Number of marketing terminals	102
Number of retail outlets	24,600

Our refineries must run crudes whose characteristics with respect to specific gravity, boiling range, sulfur content, and hydrocarbon composition match the refining facilities. The refineries must in turn be capable of using these crudes to produce the needed volumes of the specific products which match consumers' varying demand patterns. As such a system expands, it becomes increasingly important that the lowest cost transportation be used for movement of both raw materials and finished products. Coordinating these activities for Exxon Company, U.S.A. requires detailed short and long-term planning. Our vertically integrated structure permits both short-term operations and investment plans to respond swiftly to changes in supply and demand. For example, we are able to optimize investments in storage capacity at significant savings to the consumer. When we encounter a temporary disruption at the marketing end, we can generally utilize refining and terminal storage to dampen the effect. This balancing is further facilitated by exchanges of both crude and product; exchange agreements are discussed in section IV. In addition, detailed projections of product demand and assessments of market conditions are used to achieve an optimum balance between refining costs and product quality levels. To subject these operations to the spot market alone would cause chaos. The system is flexible, but not nearly flexible enough to respond to the vagaries of the spot market.

The extent and effect of vertical integration in the petroleum industry is generally misunderstood. A common misconception concerning vertical integration in the petroleum industry is the belief that only a few firms are vertically integrated. Well over 50 petroleum companies in the U.S. operate in three stages—production, refining, and marketing—and numerous others are integrated in two stages. In fact, less than 4 percent of the refining capacity in this country is operated by companies involved in the refining activity only.

There is also a misconception that only large firms are vertically integrated. Among the smallest refiners whose total capacity aggregates to less than 200,000 barrels per day, less than 5 percent of the capacity is operated by companies that are refiners only.

Another common misconception is that vertically integrated firms are totally integrated or balanced in that they produce, refine, and sell the same amounts. Large vertically integrated companies do not operate a closed system through which a symbolic drop of oil travels from the ground to the motorist's tank in a solid tube. In reality, the symbolic tube is permeable in that oil and products enter and leave the system in huge quantities at all levels.

Exhibit 1 shows the degree of crude oil self-sufficiency for the top 20 refiners ranked in terms of refinery runs. As one can see, only Getty-Skelly, ranked number 17 in refinery runs, has a self-sufficiency ratio in excess of 100 percent. Two more refiners produced two-thirds of their crude needs, and the self-sufficiency ratios for the remaining 17 firms range from 6.4 percent to 62.5 percent. The average, weighted by the volume of refinery runs, for the top eight refiners was 50 percent, which means that on average these top eight refiners had to purchase one-half of their oil for their refinery runs, ranging between 33 and 65 percent. These purchases, not only those by the top eight but by all refiners whose crude production was less than their refinery runs, plus those purchases necessary to acquire oil of the quality or location required by each refiner, mean that there is a very substantial commodity market for oil in which supply and demand interact to let the market determine price. As indicated earlier, Exxon, though it has a 62 percent self-sufficiency ratio, buys and sells more crude than it produces.

The sale of domestic refined products represents another area in which self-sufficiency is relatively low. For example, direct sales of refined products to the end-use consumer, including all industrial and utility customers, by Exxon represents only about 45 percent of our volume; the balance is sold to independent distributors and resellers. Also frequently misunderstood is the notion that large oil companies sell the majority of their petroleum products through retail outlets which they own and operate. In the case of Exxon, less than 5 percent of Exxon branded outlets are company operated service stations.

Another common misconception is the belief that vertical integration leads to a reduction in competition. This is addressed in the next section.

Competition in the Petroleum Industry

On January 21, 1975, Exxon submitted a statement to the Senate Judiciary Subcommittee on Antitrust and Monopoly demonstrating the high level of competition in the petroleum industry. That submission provided a detailed economic analysis of the structure and performance of the petroleum industry, including concentration ratios, profitability, price behavior, degree of vertical integration, ease of entry, technological progressiveness, the independence of action of the larger firms, and the economic rationale for such characteristics of the petroleum industry as exchange agreements and joint operations. Exxon believes the facts represented irrefutable evidence of the competitiveness and efficiency of firms in the petroleum industry.

A substantial body of evidence has been accumulated to test whether the various stages of the U.S. oil industry are competitive. Such evidence can be categorized by whether it pertains to industry structure, ease of entry, or rate of profit. By the usual structural tests, all phases of the industry are competitive.

Exploration and Production. Over 10,000 companies compete in oil and gas exploration and production. As can be seen in Exhibit 3, the top four crude producers in 1974 accounted for less than 26 percent of total crude production and the top eight accounted about for only 42 percent. The top four gas producers accounted for about 25 percent of total gas production and the top eight natural gas producers accounted for only about 37 percent of total production. These figures compare most favorably with four and eight firm concentration ratios for the average of total manufacturing of 39 percent and 60 percent respectively.

Moreover, it is clear that the independent sector has not only maintained but increased its share of the accelerated exploration and development activity in the U.S. According to the American Association of Petroleum Geologists, since 1969 the independent segment of the industry has drilled 9 out of every 10 new field wildcat wells and has made 75 percent of the new discoveries, although some of these were joint efforts with majors. In 1973, the *Oil and Gas Journal* reported that the independents' portion of exploratory success was up 8.9 percentage points from 1972 while the shares of both majors and drilling funds fell. The independents' success continued in 1974, as they drilled 85.2 percent of the exploratory wells and completed 78 percent of the discoveries. While independents make about 75 percent of the discoveries, they find about one-half of the oil and gas.

Similarly, independents have been increasingly more active offshore. In 1973, Secretary of the Treasury William Simon analyzed the most recent Outer Continental lease sales and concluded, "...the results show very active competition.... Independent oil companies were successful, either jointly or alone, in obtaining positions in about 68 percent of the tracts offered in the last three sales." Indeed, during the 10 federal offshore sales since 1967, nearly 150 firms (or groups of firms) submitted bids. Moreover, the number of bidders has been increasing with each sale, along with the number of bids.

Two offshore lease sales held in 1974 serve as a further illustration of the fact that smaller firms are increasing their participation offshore. In the offshore Louisiana sale, 81 separate companies entered bids, and 66 companies, or 81 percent of those entering bids, participated in high bids on one or more tracts. Of those 66 companies, 57 companies had bids accepted by the Department of the Interior. In the offshore Texas sale in 1974, 77 companies entered bids and 55 companies ultimately had bids accepted by the Department of the Interior.

Refining. On January 1, 1975, there were 133 refining companies operating 264 refineries. As can be seen in Exhibit 3, concentration ratios in refining are low both in absolute terms and relative to the average for total manufacturing. The four-firm concentration ratio in 1974 was about 31 percent and the top eight firms accounted for about 54 percent of refining capacity. These concentration ratios have been declining since 1950, and the trend accelerated after 1970.

Since 1950 there have been a number of new entrants in refining and some have grown rapidly to substantial size. Among these are American Petrofina (200,000 BPD), Amerada Hess (728,000 BPD), Coastal States Refining Co. (212,982 BPD), and Koch Industries (109,800 BPD).

The rapid growth of the new entrants along with that of the so-called "independents" has caused the already low concentration in refining to fall even lower.

Moreover, the construction of new independent refining capacity would be even greater without environmental opposition. A March, 1975 FEA analysis concludes that environmental opposition remains as a key issue affecting independents' growth. Environmental opposition has played a role in the cancellation of at least 11 proposed refineries, although other factors also contributed to the cancellations. Ninety percent of the nearly 2 million barrels per day of cancelled capacity was to have been constructed by independent refiners. FEA rules restricting refining profitability have also been a significant deterrent to entry and to expansion by all refiners—large and small alike.

Marketing. The 1973 Preliminary Staff Report of the Federal Trade Commission acknowledged that, ". . .gasoline marketing is the most competitive area of the petroleum industry and has the largest number of independent companies." There are more than 15,000 wholesalers and more than 300,000 retailers of motor gasoline, most of whom are independent businessmen.

As shown in Exhibit 3, concentration ratios in marketing are also low in absolute and relative terms. In motor gasoline sales, the top firm had 8 percent, the top four firms had 30 percent, and the top eight had 52 percent. The concentration ratios for total petroleum product sales were quite similar. Moreover, independent marketers have increased their share of U.S. retail motor gasoline sales from about 20 percent in 1968 to almost 30 percent in 1975.

In addition, plans for significant new entry into marketing are continuing to be developed. Crown Central Petroleum Corporation, a large independent refiner, is negotiating to acquire acreage in Baltimore for a new refinery and plans to market its product in that area. JOC Oil has plans to build a 200,000 barrel per day refinery in Louisiana and to open 18 direct-operated service stations in New Jersey. The Oil Shale Corporation, which was not involved in petroleum marketing in 1970, is currently negotiating to purchase over $300 million of the refining and marketing assets of Phillips Petroleum, which will add to the marketing assets that it already owns.

Once again, as with refining and exploration and production, the large number of companies, the low levels of concentration, and the continual growth of independents is hardly indicative of a market structure dominated by a few.

Transportation. In transportation, pipelines, tankers, and barges often compete with each other for carrying crude and products. Over 100 separate pipeline companies engage in interstate movement of crude oil and products. Additional pipeline companies operate intrastate, primarily as crude gathering and trunk lines. The top four pipelines account for about 34 percent and the top eight about 55 percent of total volume moved by interstate pipelines, with no single line moving more than 11 percent of total volume shipped interstate.

Pipelines move about 73 percent of the crude transported to refineries; the rest is moved by barges, tankers, trucks, and rail. The U.S. inland fleet is highly diversified with major companies owning only about 17 percent of the total tonnage and independent fleet operators owning the remainder. Trucks and rail carry only 1 percent of the crude moving to refineries.

Pipelines do not serve as a competitive advantage to the owners and do not stifle competition from others. Ownership of pipelines is dispersed among many companies. Furthermore, the vast majority of the pipelines are common carriers regulated by the Interstate Commerce Commission to insure that they are available to all on a nondiscriminatory basis. Recently Interstate Commerce Commission Chairman George M. Stafford testified that the ICC historically had received "very few complaints with respect to alleged discriminatory rates and practices of pipelines." Of more importance, he stated ". . .there is no evidence that the past or present practices of joint venture pipelines have discriminated against independent shippers."

This viewpoint is corroborated by a 1973 letter to all U.S. Senators from T.B. Medders (President of the Independent Petroleum Association of America). He urged the defeat of Senate Bill 2260 calling for divorcement of pipelines. He stated, ". . .we are not aware of any producer having difficulty selling or moving his crude oil, and we do not believe any such discrimination exists. Divestiture, we believe, could serve to increase costs for all independents and raise prices for consumers."

With the decline in domestic crude production and the increase in foreign oil imports, idle capacity in existing pipelines is becoming increasingly significant. From 1972 to 1974, there was a decline in deliveries of crude oil by pipeline to refineries of approximately 7 percent for domestic crude and 5 percent for all crude. For interstate lines alone, deliveries of crude oil and products fell 3 percent in 1974 and net income fell almost 24 percent. It is difficult to understand how it could be in the interest of pipeline owners to discriminate against certain shippers when there is excess capacity and falling profits.

Even with the recent declines in profits, there has been some criticism that owners of pipelines earn monopoly profits, and that dividends received by pipeline owners are in reality rebates which give pipeline owners a competitive advantage over other shippers. Yet the facts clearly indicate that this allegation is not true. First, under ICC regulations, all companies have equal access to interstate pipelines at the same ICC regulated tariff. Second, the 1941 Pipeline Consent Decree limits annual dividends paid to those integrated pipeline owners, based upon the pipeline's ICC valuation.

Joint Operations. Critics of the petroleum industry have often charged that joint operations in pipelines and in exploration and production have created barriers to entry. Facts have already been presented showing that neither joint operations of pipelines nor joint operations in exploration and production have hindered competition. To the contrary, in recent years joint operations have permitted both large and small operators to spread risks in offshore operations in the face of the steady upward trend of offshore acreage prices and the rank wildcat nature of some acreage offered at recent sales. From 1954 to 1973, joint operation bids went from 10 percent of all bids submitted to about 85 percent. However, contrary to what some might believe, the increased incidence of joint bidding over time has been associated with a 47 percent increase in the number of bids per tract.

With respect to pipelines, both joint interest and undivided interest lines have allowed the companies to realize large economies of scale and, thus, lower transportation costs which translate directly into lower prices for the consumers of petroleum products.

Exchange Agreements. One of the little understood characteristics of the petroleum industry is the wide use of crude and product exchange agreements. As with joint operations, many critics allege that they serve as a barrier to entry. In actuality, exchange agreements promote competition by

— facilitating entry and expansion in markets where a company does not have its own refining facilities;
— giving refiners greater flexibility in securing crude of the appropriate quality, grade, specific gravity, and sulfur content;
— reducing inventory and operating costs and supply interruptions caused by temporary shortage/surplus situations at refineries or terminals;
— reducing transportation costs.

An analysis of Exxon, U.S.A.'s crude and product exchanges for the typical year 1970 shows that Exxon has exchange agreements with both majors and non-majors. In fact, as seen in Exhibit 2, Exxon conducted the largest percentage of its exchanges, both in number and by volume, with non-majors (companies not among the largest 16 refiners).

A typical example of a crude oil exchange agreement is represented by the agreement between Exxon and Lion Oil Company, a small refiner in El Dorado, Arkansas. Lion receives approximately 2,000 barrels per day of crude in West Texas which cannot be economically transported to their El Dorado refinery. Exxon transports this crude to Baytown, Texas for utilization in Exxon's refinery there. In turn, Exxon ships to Lion 2,000 barrels per day of Exxon's East Texas crude which can be transported directly to the Lion refinery via the Mid Valley pipeline.

A typical product exchange agreement is represented by Exxon's agreement with the Thunderbird Petroleums, Inc., a small refiner and marketer in the northwest section of the U.S. Exxon receives approximately 15,000 barrels per month in refined products from Thunderbird at Cut Bank, Montana and Kevin, Montana, site of their refinery capacity. Exxon returns the 15,000 barrels per month at Portland, Oregon and Spokane, Washington, where Thunderbird markets in competition with Exxon, but where they do not have economical access to their refinery capacity.

If exchanges were abolished, they would be replaced by an increase in spot buying and selling and some increase in cross-hauling until a refiner-marketer could justify building capacity in, or transportation facilities to, new market areas.

It is clear that both crude and product exchanges are implemented for a variety of valid economic reasons, and not for the exclusion of any class of competitors, that they do not result in any anticompetitive behavior, and that their abolition would reduce competition, increase consumer prices, and increase the probability of localized product shortages.

Independence of Action. Another measure of competitiveness of the petroleum industry is the diversity of the firms in the industry. There are numerous large and small firms and each of these pursue independent strategies based upon individual strengths, weaknesses, assessments, and policies. Of course, it should not be surprising that many firms do react in similar fashion when confronted with similar government regulations, economic conditions, and market forces.

Nevertheless, there are significant differences—even among the major petroleum companies in crude self-sufficiency ratios; in capital expenditure programs and their geographical and functional allocation; in degree of investment overseas; in methods of financing expenditures; and in degree of participation in natural gas, chemicals, and other energy fields. The results of these differences are manifested in divergencies and changes both in net income and in relative ranking in market shares among the companies in each function.

Because of this diversity, that which is attractive to one firm is often not attractive to another. This diversity of interest and perceived opportunities is a vital force for competition within the petroleum industry and should be encouraged—not discouraged as would be the case with vertical dismemberment.

Profits. A final measure of competition in the petroleum industry is profitability, which is really the key to determining if these other characteristics are effecting true competition in the industry. Historically, profits in the petroleum industry have been about equal to the average for all of U.S. manufacturing. For instance, from 1967-1973, the rate of return on net worth for the petroleum industry was 12.3 percent, exactly equal to the average for all U.S. manufacturing.

In 1973 and 1974, profits did increase substantially in the petroleum industry, but profits in essentially all industries increased in that same period. Much of the increase in petroleum was due to recovery from very poor profit performance in Europe and elsewhere overseas combined with dollar devaluation, inventory profits, and higher chemical earnings.

Even with the well-publicized increases, the rate of return on net worth in the petroleum industry in 1973 was 15.6 percent, only slightly higher than the 14.8 percent average return for total U.S. manufacturing. In 1974, the average rate of return on net worth for the petroleum industry was 19.9

percent, but once again profits included a substantial amount of inventory profits.

In the first nine months of 1975 there has been a substantial decline in oil industry profits. Sales volumes declines because of conservation and the slowdown in the economy. Also, the petroleum industry has experienced higher than average inflation for the materials and equipment used by the industry. Finally, the decline in profits also has been caused substantially by the loss of the depletion allowance, lower foreign earnings, the decline of chemical earnings, the disappearance of inventory profits, and the detrimental effects of the Federal Energy Administration's allocation and entitlements program.

An analysis of 25 leading oil companies showed that the rate of return on net worth for the first six months of 1975 fell to an annual rate of 11.9 percent. This was a 41 percent decline from the rate of return of 20.1 percent for these same companies in the first six months of 1974. This 11.9 percent rate of return on net worth is less than the average for petroleum and all other industry over the period 1967-1973. It is also interesting to note that this 41 percent decline would fully offset a 69 percent increase in the previous year. (100 percent + 69 percent = 169 percent; 169 percent −.41(169) percent = 100 percent) Profits in the petroleum industry in 1974 actually increased by 41 percent.

Available nine months earnings reports for 1975 show this downward trend continuing. Actual net income for the top 20 oil companies is down over 21 percent from the same period in 1974.

This loss of cash flow and the resulting lower rate of return, combined with the much higher costs of energy development, has already caused a slowdown in the drilling activity in the U.S. and generally slowed the drive to build more facilities for processing and transporting energy supplies. After a 20 percent increase in well completions in 1974 over 1973 and a 22 percent increase in the first quarter of 1975 compared with first quarter 1974, there was a decline of 6 percent in the number of wells drilled in the second quarter of 1975 compared with the first quarter.

Most estimates project the petroleum industry capital expenditures over the next decade will have to increase by 200 to 400 percent. Higher rates of return and profits will be needed to supply petroleum companies with the means to develop U.S. energy supplies and reduce the critical trend of increasing reliance upon foreign energy supplies.

The Impact of Petroleum Industry Dismemberment

The dismemberment of the nation's twenty largest vertically integrated petroleum companies into separate functional entitites would reduce the nation's domestic energy supplies, significantly increase the costs of energy to consumers, frustrate, if not reverse, attempts to recover from the recession, reduce competition in the petroleum industry, and have harmful effects upon the many millions of investors whose savings, incomes and pensions depend upon the returns from stocks and bonds.

Effects upon U.S. energy development and the U.S. economy. If dismemberment legislation were to be passed, there would be significant transitory effects which would have both long and short-run implications. During the time between the passage of such legislation and the time it is implemented, the industry would be operating in an environment of total uncertainty. The immediate impact of such legislation on most affected companies in our judgement would be to suspend capital spending programs and focus top management attention on developing its dismemberment plan and on reevaluating and restructuring short and long-range investment plans.

An example of the seriousness and magnitude of the problem is the necessary restructuring of petroleum company debt that would be required. The debt burden of 20 major petroleum companies in 1974 was already $22 billion, and has grown significantly since then to finance the increased capital expenditures of 1975. This debt burden is largely secured by corporate guarantees from financially sound integrated firms. Since the firms creating the debt would no longer have the same earnings and asset bases, new guarantor arrangements would have to be established. This would be complicated by the fact that, in many cases, covenants in existing indenture agreements prohibit the disposal of the corporation's underlying assets.

A prime example in this category is the Trans-Alaskan Pipeline where new owners, operators, and financial backers would have to be found for the over $6 billion investment, which is the largest single private enterprise venture in our country's history. Construction on TAPS and Prudhoe Bay may well be shut down indefinitely until new arrangements could be made, if they can be made at all. While TAPS is an obvious example, similar considerations would be involved with regard to the entire debt structure of over $22 billion. The slowdown or suspensions of other investment plans and projects would seriously impact the nation's energy capabilities in many ways. For example, major refinery expansions could be suspended until contracts for secure and economic raw material supplies could be obtained on one end and product sales agreements signed on the other end. In addition, legislation of this type could create additional incentives for capacity expansions to swing to foreign producing countries that are actively industrializing. Domestic self-sufficiency in refining would obviously suffer.

Capital investments for synthetics development such as the liquefaction and gasification of coal most likely would be curtailed pending the development of a clear incentive in one or more of the dismembered functions. Indeed, it is not certain whether the definition of synthetics manufacturing would place it in the production function or refining function, or neither. These curtailments would be in addition to those resulting from increased capital costs and political uncertainties which would reduce the overall spending level on alternate energy sources.

The resulting effects of this curtailed investment on business plans and employment throughout the economy would slow, if not reverse, the current economic recovery. The petroleum industry and related support industries have been important in minimizing the depth of the recession and in accelerating economic recovery. For instance, capital and exploration expenditures in the petroleum industry alone (not including expenditures for chemicals, in other energy industries, and in energy support industries) amounted to about 11 percent of total business fixed investment expenditures in 1974 and will certainly increase significantly in 1975. Obviously, any reduction in activity in the petroleum industry would doubtless have important depressing effects upon the overall economy.

Exxon estimates that the direct effect of even a 50 percent reduction in petroleum industry capital expenditures next year would increase unemployment by approximately 300,000 workers in 1976 and reduce the U.S. economic growth rate in real GNP by 14 percent from 5.8 percent to 5.0 percent. The indirect effect of this reduction in investments is much more difficult to estimate. Nevertheless, it is likely that, because of the key role the petroleum industry is playing in the nation's economic recovery, this drastic reduction in expenditures will have a depressing psychological effect upon consumers and other industries. If this psychological effect simply leads to other industries holding new capital expenditure plans in 1976 and 1977 to 1975 levels, then this would add as many as 470,000 people to the ranks of the unemployed in 1976. Moreover, the combined effects of no increase in capital expenditure plans in 1976 and 1977 over the 1975 levels would mean an increase in unemployment of over 1 million people in 1977 beyond that which would otherwise prevail.

The rate of growth in real GNP in 1976 would be reduced by over 25 percent to 4.3 percent instead of 5.8 percent. In 1977 the growth rate would be further cut by almost one-half to 2.4 percent. In current dollar terms, the total loss in GNP over this two-year period would be almost $85 billion.

The long-run implications of this reduction in energy development investment would have even more significant detrimental effects upon the U.S. economy. The U.S. is currently experiencing declining capability in domestic energy production and the only hope of reversing this trend is rapid and sizeable investments on a broad front. The lead times for bringing forward resources are long and subject to many potential disruptions. For example, the lead time for bringing offshore crude to initial production is 4 to 8 years, and time to peak production is frequently several years longer. Coal mines require about five years, lead time. Any actions taken today, which have the effect of temporarily curtailing industry spending, have a direct adverse effect upon the availability of energy supplies in future years. Technology, manpower, and capital resources in the petroleum industry are stretched to the limit in today's environment. Lost opportunities and lost time cannot be recovered through efforts in the future.

Should dismemberment legislation delay the development of new energy resources for its three year implementation period—which is not only quite possible but may even be optimistic in terms of the time involved—available domestic energy supplies would be reduced by 2.5 million barrels per day of oil equivalent in 1980 and by about 4.0 million barrels per day of oil equivalent by 1985. At a minimum, this will result in a reduction in GNP in the U.S. economy averaging $9 billion per year (in 1975 dollars) for the period 1975-1985. Assuming the best, the U.S. economy will replace this lost production with imports. At today's costs this will add over $100 billion to our balance of payments drain over this ten-year period.

This reduction in value added in the U.S. will be reflected directly in either lower consumption and lower investment by Americans in the U.S. or higher imports and increased balance of payments deficits averaging over $10 billion per year. This will result in a lower standard of living for the U.S.

Increased costs to the consumer. Dismemberment legislation would eliminate all of the efficiencies of vertical integration which were discussed in the sections above and which brought cost savings to the consumer. Specifically, dismemberment would remove the advantages which vertical integration presents in terms of:

1) facilitating complementary and highly interdependent investments and the optimal sharing of risks resulting from increased assurance of crude supplies and product outlets;
2) optimal scheduling of inventories and production;
3) maximum incentive for technological innovation;
4) ability to raise the tremendous amounts of capital required to pursue the energy goals of the nation; and

5) ability to minimize intermediate buying, selling, advertising, and communications costs, as well as the costs involving additional management and staffs for such activities as planning, law, treasury, accounting, research, employee relations, public affairs, and other activities.

Exxon has estimated that all of these dismemberment-caused inefficiencies on an industrywide basis would add over $5 billion (in 1975 dollars) to the recurring annual costs borne by the U.S. consumer. These increased costs include increased industry capital costs; reductions in transportation economies of scale; less productive manpower utilization; an increased level of required working capital; an increased level of required inventories and tankage; and increased costs caused by higher business risks in refining and marketing. Adding this $5 billion to the $9 billion a year (in 1975 dollars) of lost consumer purchasing power, which results from the need for increased foreign oil imports, results in a total recurring annual cost of dismemberment legislation which will be borne by the consumer of over $14 billion (in 1975 dollars).

Effects upon the U.S. stock markets and bond markets. The detrimental effects upon the economy of reduced petroleum industry activity and the possibility of energy shortages, combined with valid concerns of the investment community regarding possible further dismemberment moves by Congress against other vertically integrated industries, would create the one thing capital markets abhor the most—uncertainty. This could trigger a general downturn in stock values and bond values. The investments of millions of individual Americans with values in the hundreds of billions of dollars would be endangered. The 1970 New York Stock Exchange Census indicated that there was an overall total of 31 million individual shareholders with investments valued at over $1 trillion. These do not even include the millions of citizens who have invested through such institutional accounts as mutual funds, mutual insurance companies, pension funds for state and municipal employees, and union and private company retirement plans.

Exxon is owned by 707,000 shareholders, 450,000 of which own less than 100 shares. Looking at a broader group, of the 2.3 million individual shareholders of the six largest petroleum companies, 46 percent are retired individuals with an average income of about $14,000 a year. However, it is estimated that 11.7 million people hold investments in these companies indirectly through such institutions as mutual funds and pension funds. This totals 14 million people. Additionally, 91 colleges and universities and about 1,000 charitable and educational foundations are shareholders of the six largest oil companies. These figures should give one a good idea of both the large numbers and the profile of the many millions of investors that could be affected by dismemberment legislation.

Dismemberment legislation also would have a depressing effect upon the bond market and the millions of corporate bondholders. The debt burden of 20 major petroleum companies in 1974 was $22 billion. The uncertainties regarding source of payment of this debt would immediately increase the risk and reduce the desirability of these bonds for existing bondholders and would inhibit new capital availability and flow to the companies. There would be serious legal questions about who would be responsible for these debts. In the case of the Trans-Alaska Pipeline, the terms of the federal permits for the project require parent company guarantees of compliance throughout the project life. Yet the companies required by law to be responsible for the endeavor would be forbidden by law to participate in the pipeline. Moreover, it seems certain that companies with no financial background and questionable near term earnings could not raise the capital to provide this much needed transportation facility. The resulting chaos may well render private industry incapable of continuing construction on the line indefinitely.

This uncertainty about the $22 billion or more in existing debt for 20 major oil companies and the uncertainty about any new debt offerings would have an important influence on capital markets in general. Any reluctance to purchase new petroleum company bonds or any degree of selling of existing petroleum company bonds on the open market by investors who believed that the interest rate no longer compensated for the risk involved would drive up interest rates, cause capital losses for existing bondholders of all companies, and increase the costs of all capital to all companies. The degree of uncertainty that will pervade the economy and the likelihood of a slump will mean that it will be unlikely that the investors who dispose of petroleum company bonds will be willing to buy the debt instruments and equities of other corporations. Even if they do reinvest, a higher return will likely be demanded for risk compensation. Once again, this would increase costs to the consumer, not only in the petroleum industry but in the economy in general.

Effect upon competition in the petroleum industry. As discussed above the structure of the oil industry in terms of number, size, and diversity of companies is a strongly pro-competitive factor. Dismemberment legislation would create a larger total number of companies, but at each of the levels—producing, refining, transportation, and marketing—there would be the same number of companies as there are now, or possibly less. Competition would not be increased since there is no inter-functional competition, only competition within like lines of business. Moreover, there would be a tendency to eliminate the diversity of interest which together with low concentration ratios and ease of entry constitute the strongly competitive factors in the petroleum industry.

Dismemberment legislation would complicate the ease of entry which has always characterized the petroleum industry. Companies which have started in one function and wish to integrate into others where it appears more efficient to do so would be barred from competing in such fields. In addition, many firms which have capacities or sales close to the arbitrarily established criteria for dismemberment would not have an incentive to compete and obtain a greater share of the market since the reward for success is being forced to dismemberment. Others which slightly exceed the criteria may well reduce capacity rather than dismember. If the companies themselves tried to create such a structure of limiting markets and erecting barriers to entry as is created by dismemberment legislation, they would be charged with a violation of antitrust laws. It follows that the legislative dismemberment vehicle that does the same thing must also be anti-competitive. It is counter to the objectives of antitrust law. It will result in less competition, fewer supplies, and higher prices.

As has been discussed in detail in the sections above, the necessities of assured supplies of both raw materials and products, and the complexities of short and long-term planning preclude the extensive use of either spot market transactions or short-term contracts for the various interfaces between finding the raw material and delivering the product to the final consumer. Thus, after the initial uncertainties and problems of the transition period, most likely long-term contracts would be used to replace corporate ties with contractual relationships. Although long-term contracts are far from perfect substitutes for vertical integration, they would assure that there would be no essential change in the economic structure of the petroleum industry. The only significant change would be the increased costs to the consumer that result from the increased risk and uncertainty, the increased inefficiencies and duplication, and the other increased costs of long-term contracts compared with vertical integration.

Summary of the effects of dismemberment legislation. It has been shown in this section that dismemberment legislation would significantly increase costs to the consumer. It would not increase competition and, in fact, would create barriers to entry. Dismemberment legislation could very well have damaging effects upon the many millions of investors whose savings, incomes, and pensions depend upon the returns from stocks and bonds. Finally, it would have detrimental effects upon energy development in the

U.S. and, as a result, short-term and long-term detrimental effects upon the U.S. economy.

International Aspects of Dismemberment

It has been claimed that the international oil companies—whether or not intending to—support OPEC prices by the way in which they draw supplies from individual OPEC countries. An additional claim is that the companies have profited from the drastic increases in prices imposed by OPEC and therefore have not taken effective steps to oppose these price changes. These charges are being used as justification for dismemberment legislation. Dismemberment of the integrated companies into separate functional organizations would, according to the claimants, weaken OPEC's ability to control crude prices.

The facts do not support these claims. Dismemberment would not be effective in putting pressure on OPEC because it would not deal with the factors that provide OPEC with its market strength. In fact, dismemberment would strengthen OPEC by slowing down the development of alternative energy supplies. Finally, this so-called remedy would have an adverse impact on the U.S. consumer, both with respect to prices and security of supply.

The inescapable fact is that the consuming nations continue to be heavily dependent upon OPEC nations for oil. Despite efforts to conserve, this dependency is growing. On the other hand, the cumulative revenues from oil sales of the OPEC nations as a group already exceed their capacity to absorb the goods and services which they can buy with those revenues. Moreover, and this is equally important, because of its large productive capacity and its relatively low needs for revenue, one country alone, Saudi Arabia, holds the key to control of the general level of prices. It is within its ability to absorb major production decreases if, in its judgment, these become necessary to sustain prices. What makes high prices for OPEC oil possible, then, are two things: first, the fact that the consuming nations continue to be dependent upon OPEC oil while OPEC nations collectively are generating surplus revenues from the sale of that oil and, second, the peculiar distribution of reserves and productive capacity within OPEC. This situation would in no way be altered if the oil companies were to be stripped of their foreign producing operations.

What role have the international companies played with respect to OPEC production over the past 18 months? To begin with, there are at least 25 large international oil suppliers dealing with these OPEC countries, and several times this number of smaller companies—not just the seven of popular myth. These companies have effectively provided supply continuity to major importing countries during a period when the companies' producing interests have been continuously whittled away. These companies have also actively bought and traded oil in the open market such that OPEC countries attempting to price their oil exports above parity with benchmark crudes have been forced to either lower their prices or accept reduced export volumes. In performing these functions, the companies have dealt independently with the producing governments and with each other—there has never been any credible evidence of collusion throughout this very turbulent period.

The question has been raised as to why the companies do not shift their lifting patterns to place maximum pressure on the more vulnerable OPEC members and thereby force a major price reduction. To begin with, such action, to be effective, would have to be joint action by the companies, which is clearly against U.S. law. Moreover, it would quite probably further solidify OPEC against what would be perceived as a clear external threat. In our view, collective action to force a price reduction would not be commercially appropriate and would have to be undertaken by governments, if undertaken at all. We believe that it would be ill-advised, would offer little chance of success, and would risk the type of confrontation that most consuming governments are trying to avoid.

What of the related charge that the oil companies do nothing to resist higher OPEC prices because they themselves profit from them? The record will show that the companies have resisted the price increases and that oil company producing profits from OPEC crude have decreased significantly over the period of OPEC price increases. In Exxon's case, producing profits on a representative light Arabian crude were in the 35¢ per barrel range early in 1973 when crude market prices were about $2 per barrel. Today, with Arabian light crude selling at $11.50 per barrel, our comparable producing profits have been reduced to less than 25¢per barrel. The record will also show at least in the case of Exxon that its worldwide profitability is currently running at pre-crisis levels.

While dismemberment legislation can be expected to have little impact on OPEC, it would prove costly to the U.S. consumer. Our assessment of the outlook for energy demand and supply in the U.S. suggests that the need for imported oil supplies must *increase* in the future if an adequate level of economic growth is to be sustained. By 1985, it is projected that U.S. import requirements, under current conditions, will grow to about 50 percent of U.S. oil demand. In this environment, weakening U.S. based international oil companies, either through domestic or international dismemberment, would seriously increase U.S. exposure to supply disruption and would increase the cost of foreign oil supplies.

Supply during periods of shortage is a complex international logistics problem. The Church Committee, along with recognized international experts outside the industry, have praised the manner in which the international companies operated to blunt the actions of the embargoing producing countries and to fairly allocate existing supplies during the cutoff period two years ago.

Looking toward future collective security planning, the important interface with governments in the implementation of the emergency sharing and alternate energy development programs being developed by the International Energy Agency, would be lost through dismemberment. This responsibility is currently being discharged by the existing international companies and there is no assurance that many new and smaller companies oriented toward single functions could successfully fill this important role.

Another unintended result of dismemberment would be to increase the cost of foreign oil supplies. The integrated supply systems that have been developed by a great many international companies over a period of many years achieve significant cost savings in the international environment of rapidly fluctuating demand for petroleum products. This is accomplished by:

— continuously buying and selling crude oil in the open market, and adjusting owned crude oil production so as to come up with a composite crude stream that most economically meets each company's aggregate worldwide demand pattern,

— utilizing the lowest cost combination of refining and transportation facilities,

— permitting quick and effective adjustment for the inevitable supply upsets caused by factors such as weather or facilities failures.

Possibly the most serious unintended effect of dismemberment would be the slowing down of the development of international energy supply alternatives to OPEC. Dismemberment would result in an extended period of reorganization, adjustment, and uncertainty for U.S. companies. There would be fewer investors willing to take the greater risks and the costs would be higher. The result would be a drastically diminished role for U.S. companies in worldwide resource development.

As a last observation, it should be noted that dismemberment could have significant implications for our relations with other nations. The international flow of capital has been an important factor in the economic development of all countries during the postwar period. In recognition of this, the principle of free international investment flow has heretofore been strongly supported by the U.S. Forced functional dismemberment of U.S. operations would impact on a number of foreign-owned companies operating in the U.S. This action would signal a major shift in U.S. policy on international investment. It would undoubtedly cause foreign governments to reconsider their policies on accepting U.S. investments, and could impact on a wide range of U.S. investments overseas.

Moreover, an attempt by the United States to apply American dismemberment requirements to foreign operations of American-based companies could be seen as interference in the internal affairs of other countries. Other governments have long resented any extraterritorial application of U.S. laws, and they can be expected to be particularly sensitive in the case of an industry which is so vital to their economic well-being.

Summary and Conclusions

In considering legislation that could markedly alter the structure of the petroleum industry, it is important that Congress understand the current operations of the petroleum companies, the values that these operations provide to the consumer, and the risks to both the economy and to long-term energy supply that are inherent in any massive restructuring of the industry. In this submission, factual evidence has been presented that vertical integration is a logical and natural structure for many industries including petroleum. Such integration permits higher utilization of capital intensive facilities and improved coordination of complex operations that assure a smooth flow of petroleum products to the consumer at the lowest prices.

The extent of vertical integration in the petroleum industry is greatly misunderstood. Evidence was presented that many firms both large and small are integrated into more than one segment of the industry, and that those that are integrated are far from being fully integrated in all functions. As a natural outgrowth, competition has been fostered in all segments of the business by the ease with which a company operating in one segment can integrate into another. Passage of dismemberment legislation would actually reduce competition by raising barriers to entry, rather than increasing competition as claimed by some. The net effect on the consumer would be higher prices for energy.

Of even more concern than the higher energy costs would be the impact that dismemberment legislation would have on the economy and the nation's domestic energy sufficiency. Passage of such legislation would result in an immediate curtailment of energy investments that would adversely impact economic recovery and result in continued high unemployment. Curtailment of near term petroleum industry investments would lead to increased dependence on imports of foreign petroleum. These higher levels of imports would reduce the purchasing power and standard of living for U.S. families while simultaneously increasing the vulnerability of the nation to another embargo of foreign crude. This increased dependence upon imports could strengthen OPEC, not weaken it. Finally, the detrimental effects on the economy combined with valid concerns in the investment community about the results and uncertainties of dismemberment would endanger the savings of the investing public.

In conclusion, the evidence is convincing that legislation to alter the structure of the petroleum industry will be detrimental to the nation and the consumer. The major challenge facing the Congress and the industry is developing and implementing those policies and plans which will move the nation toward increased domestic energy sufficiency. Only by addressing our attention and energies to these critical issues now can we hope to insure the long term well-being of the nation.

EXHIBIT 1

CRUDE OIL SELF-SUFFICIENCY RATIOS

Rank by Refinery Runs 1973	Name	1973 Self-Sufficiency Ratio
1	Texaco	67.8
2	Exxon	62.5
3	Shell	51.6
4	Soind	48.8
5	Socal	47.1
6	Gulf	53.1
7	Mobil	35.7
8	ARCO	47.9
9	Sun	53.2
10	Union	56.7
11	Sohio	6.4
12	Phillips	31.0
13	Ashland	7.1
14	Continental	58.8
15	Cities Service	53.4
16	Marathon	67.2
17	Getty/Skelly	130.7
18	Champlin	30.6
19	Petrofina	10.8
20	Murphy	15.6

Source: Chase Manhattan Bank.

EXHIBIT 2

1970 EXXON USA EXCHANGES

	With Others in Top 8 Largest Majors	With 9th thru 16th Largest Majors	With Non-Majors
Crude Exchanges			
% by Volume	35.5%	26.8%	37.3%
% by Number	23.1%	17.1%	59.8%
Product Exchanges			
% by Volume	35.2%	16.6%	48.2%
% by Number	37.4%	22.2%	40.4%

EXHIBIT 3

COMPARISON OF CONCENTRATION RATIOS OVER TIME

	Source	Top 4	Top 8
Production (Crude, Condensate and N.G.L.)			
1955	(1)	18.1	30.3
1974	(2)	26.0	41.7
Refining Capacity			
1950	(3)	33	56
1974	(4)	31	54
Oil Pipelines (Interstate—Barrel Miles) By Ownership			
1951	(5)	44	71
1974	(6)	34	55
Motor Gasoline Sales			
1954	(7)	31	54
1974	(8)	30	52
Natural Gas Production			
1955	(9)	21.7	33.1
1970	(10)	25.2	39.1
1974	(11)	24.7	37.2

Sources:

(1) Company Annual Reports and Statistical Supplements; *National Petroleum News,* Mid-May Issue, 1956; U.S. Bureau of Mines (U.S. Total).

(2) 1974 Company Annual Reports and Statistical Supplements; *National Petroleum News,* Mid-May Issue, 1975; *1974 Annual Review of California Crude Oil Petroleum,* Conservation Committee on California Oil Producers (City of Long Beach Data); U.S. Bureau of Mines (U.S. Total).

(3) Petroleum Administration for Defense, June 11, 1951.

(4) National Petroleum Refiners Association, "U.S. Refining Capacity," September 23, 1975.

(5) Interstate Commerce Commission, *Moody's Industrial manual;* from Barbara M. Loveless, "The Interstate Oil Pipeline Market", API, March, 1976.

(6) *Oil and Gas Journal,* August 18, 1975; Moody's Industrial Manual, 1975.

(7) De Chazeau and Kahn, *Integration and Competition in the Petroleum Industry,* 1959, pp. 30-31.

(8) Lundberg Survey, Inc., copyright by Dan Lundberg, 1975. Reprinted in *National Petroleum News,* Mid-May Issue, 1975.

(9) 1955 Company Annual Reports & Statistical Supplements; *Moody's Industrial Manual;* Bureau of Mines (U.S. Total).

(10) 1970 Company Annual Reports & Statistical Supplements; 10-K Reports; Bureau of Mines (U.S. Total).

(11) 1974 Company Annual Reports & Statistical Supplements; Bureau of Mines (U.S. Total).

3

A Common Form of Industrial Organization

TESTIMONY BY ANNON M. CARD, SENIOR VICE PRESIDENT, TEXACO, BEFORE THE SUBCOMMITTEE ON ANTITRUST AND MONOPOLY, COMMITTEE ON THE JUDICIARY, UNITED STATES SENATE — NOVEMBER 19, 1975.

Summary: Vertical integration is a common form of industrial organization. It is a form of diversification for reducing risks, enabling efficient coordination of production operations; as such, it results in reduced transaction costs, coordination of research efforts at all levels and reduced capital costs. No company is fully integrated; concentration ratios are low and profits have not been excessive. Divestiture would curtail energy investment at a time when it is most needed, raise capital costs, adversely affect millions of stockholders and put U.S. oil companies at a competitive disadvantage with foreign integrated companies.

Annon M. Card speaks . . .

Vertical integration, the linking of successive stages in the production and distribution of a particular product, is a common form of industrial organization in the United States. It has contributed substantially to economic growth by increasing productivity and efficient use of the nation's resources. Industries such as steel, aluminum, automobile manufacturing, wine and paper products, as well as petroleum, are vertically integrated.

— Steel companies own and mine mineral deposits, transport iron ore, smelt it and sell products under their own brand names.

— Aluminum companies mine bauxite, transport it, produce alumina from the ore, generate the power to produce the alumina, and manufacture and sell, in some cases under their own brand names, consumer, as well as industrial goods.

— Automobile manufacturers design, tool, manufacture, and assemble, franchise dealerships under their own brand names, provide financing for purchase and manufacture spare parts.

— Wine producers own vineyards, aging and storage facilities, and bottling plants, and retail under their own brand name.

— Paper companies own timber supplies and mill and sell consumer and industrial products under their own names.

The degree of integration that a particular firm achieves in an industry is dependent upon the cost reduction and other benefits it will obtain from vertical integration. Vertical integration is a form of diversification, whereby a firm is able to reduce risks and achieve economies through investment in other functional levels. Because of the uncertainties of exploration and the large capital expenditures required for development and production, there are many joint producing and exploratory ventures in the petroleum industry, particularly among the smaller companies. Obviously these are natural ways to spread risk. Similarly, in other areas of operation, risks can be reduced by vertical integration. For example, investment in a large, efficient refinery would be riskier to undertake without a guarantee of long-term crude supplies.

Integration Is Result of Industry Characteristics

The petroleum industry contains integrated companies of all sizes. Of 131 refining companies listed in 1975, approximately 62 have chosen to integrate to some degree in the three levels of producing, refining, and marketing. These 62 companies account for 95 percent of refining capacity. Thus, it is clear that virtually all refining capacity is to some degree integrated into producing and marketing and that the benefits of integration are available to all size companies.

There are many sound economic reasons for the extent of vertical integration in the petroleum industry:

1. The physical characteristics of crude oil and products make storage and transportation difficult and expensive. Vertical integration permits efficiencies in day-to-day operations, particularly with respect to supply and distribution problems involving producing, refining, and marketing. Integrated companies, with their intimate knowledge of all

phases of the industry, are well-equipped to coordinate production operations, refinery runs and yield patterns and inventories in order to maximize operating efficiencies.

2. The high degree of capital intensity at every level makes idle capacity very expensive and creates the need to operate close to full capacity at all functional levels. For example, the cost of constructing a new 200,000 BPD fuels refinery in the United States with full cracking facilities, which meets all ecological requirements, is now over $1 billion. Such a refinery when operated at full capacity can achieve considerable savings from economies of scale. Operating at less than full capacity dissipates the economic savings of large scale very rapidly. Integrated operations reduce the risk of running individual facilities at less than economic levels.
3. Vertical integration enables oil companies to coordinate capital investments in different segments of the petroleum industry. Investments in producing, refining, transportation and marketing facilities can be better balanced with one another, thereby maximizing capacity utilization and minimizing costs. The logistics of petroleum movements can also be enhanced by constructing new facilities in advantageous locations. Since each operating department has access to the company's overall requirements and plans, the possibility of unsound investment decisions is reduced.
4. In the absence of vertical integration, producing companies would have to sell to refining companies, which would, in turn, sell to marketing companies. Each non-integrated company would have to maintain sufficiently large purchasing organizations to buy from the previous stage of operation, and large sales organizations to sell to the succeeding level of operation. However, through vertical integration companies can eliminate much of these buying, selling and promotional costs since oil remains within the company moving operationally from one department to another. As a result, savings in cost are generated for the vertically integrated companies to pass on to consumers.
5. Vertical integration enhances a petroleum company's research and development capabilities. A major advantage is the ability to coordinate research efforts at all levels of operations to provide consumers with improved products at a minimum of time and cost. In a vertically integrated oil company, research is continually underway which will eventually benefit more than one functional level. For example, the development of a new product line or modification of an existing one

could affect both refining and sales functions. If different companies were required to perform these operations, each would be required to conduct separate, uncoordinated and overlapping research and development programs. Research would be less effective, with longer lead times required to achieve results.

6. As members of one of the most capital-intensive industries in the world, oil companies invest billions of dollars every year to meet the growing demand for energy. Because of their efficiency and the lower risk attained through integration, vertically integrated companies of all sizes are in a much stronger position to generate and attract the large funds needed for this job than would be the less efficient firms that would result from divestiture.

The Degree of Integration in the Petroleum Industry

Although most firms in the industry are vertically integrated to some degree, their ability to supply all their needs at any particular functional level is far from complete. An indicator of the degree of integration into production for the domestic refiner is its crude self-sufficiency index. This index is the ratio of net crude production to operating-refining capacity. For the latest year in which data are available (1973), the self-sufficiency index for the eight largest domestic refiners was slightly less than 50 percent. This means that the largest refiners were able to supply slightly less than one-half of their domestic refining requirements from their own domestic net crude production.

Another indication of incomplete integration is that many companies have inadequate refining capacity to meet their marketing needs and have to purchase products or arrange for processing of crude at other refineries. These processing arrangements are beneficial to the processor who is able to utilize its refining capacity at more efficient rates.

In regard to transportation facilities, most, if not all companies have to rely to some extent on others to move products or crude to market. The Interstate Commerce Commission (ICC) requires interstate pipelines as common carriers to make space available on a pro-rata basis to all shippers.

Integration into marketing is also partial for most companies. The larger companies generally have relied heavily on independent businessmen to market their gasoline and other products. By contrast, many of the so-called "independent" companies retain complete control over their service stations by operating them on a salary basis.

Considering the above facts, the idea that companies generally run their operations as entirely closed systems is fallacious. Companies depend to varying extents on others for crude oil, for pipeline and tanker transportation, for additional refining capacity, and, especially in the case of the largest companies, for the marketing of their products.

Integration Transfers No Monopoly Power

Vertical integration by itself does not confer monopoly power on any company. For example, few enterprises are as vertically integrated as the farmer selling produce on the roadside, yet he has no monopoly or market power even though he controls production from the farm to the consumer.

Moreover, in any event, there is no monopoly or market power at any level in the petroleum industry to be allegedly transferred to another by vertical integration.

Production

According to the Federal Trade Commission Staff's statistics, the level of concentration in the production of crude oil is far below that of other mineral extraction industries, as well as the average for all U.S. manufacturing industries. The Staff's January 1974 report entitled "Trends in the Energy Sector of the U.S. Economy" showed that in 1970 (the most recent year for which data are reported), the net production of the four largest crude oil producers accounted for only 27.1 percent of domestic production, while the four-firm weighted concentration ratio for all manufacturing industries was 40.1 percent. Yet, as the Staff noted, "more relevantly" the concentration ratios in crude oil production "may be compared to other mineral extraction industries such as lead and zinc mining, iron ore mining, copper mining, sulfur mining and gold mining. These industries have substantially higher four-firm concentration ratios ranging from 47.0 to 79.9 percent." Hence, the Staff concluded that "crude oil production appears to be relatively unconcentrated."

Joint ventures have been a common form of organization in the bidding for and development of offshore oil properties. By sharing the extremely high risks of offshore exploration, the bidders have been able to decrease the risk borne by each company. Joint ventures reduce risks for both large and small companies by enabling them to diversify risks, rather than "putting all their eggs in one basket." This has increased competition for leases and promoted the development of the offshore areas from which the United States will have to obtain much of its future oil production. Recently, the Department of the Interior has ruled that companies producing more than 1.6 million barrels per day of oil and gas would not be allowed to bid together. Such punitive rules will surely affect the development of the extremely risky offshore properties in the future.

Transportation

The economies of scale inherent in petroleum pipelines, and the huge investment and enormous risk needed to achieve such economies, frequently mandate joint ownership. The huge investment required for even a small, single-owner pipeline can be prohibitive; yet with a number of partners, the cost of a much larger pipeline will be significantly less for each individual owner. Naturally, this also lowers the risk in such a large capital outlay. Such large scale pipelines reduce transportation costs and consequently increase competition in the geographic regions they serve.

Although Interstate Commerce Commission regulations require that all shippers be treated as equals, non-owner shippers are, so to speak, more equal than owner shippers. While all shippers are guaranteed access to pipeline transportation, it is the owners who enter into throughput agreements, who bear the huge risks inherent in the construction and operation of the pipeline and who pay the same rates to ship over the lines as non-owners. For example, Explorer Pipeline has found that it cannot maintain optimal throughput. In this case, it is those who have guaranteed that the pipeline will pay its debts—the owners—who must bear the burden imposed by changed circumstances.

Marketing

The marketing phase of the industry is and will be increasingly competitive. Indeed, there are more than 15,000 wholesalers of petroleum products and more than 300,000 gasoline retailers in the country. Further, the August 27, 1973 Staff Analysis of the Office of the Energy Administrator, Department of the Treasury, confirmed the conclusions of most industry observers that "the independents' position in the market has strengthened at the expense of the majors." In 1967, the "independents" selling unbranded or private brand gasoline sold 19 percent of the nation's gasoline. This figure has gone up steadily—to 28 percent in 1972, 29 percent in 1973, and over 30 percent in 1974.

Profits

A primary indicator of monopoly or even market power is excessive profit over the long-term. The rate of return on net worth, as reported by First National City Bank, for the period 1965-1974, was 13.1 percent for the petroleum industry, while for all manufacturing, it was 13.0 percent. According to data based on studies by the American Petroleum Institute, the rate of return on net worth of the nation's 25 leading oil companies during January-September 1975 averaged only 12.5 percent. Considering the risk, the rate of return in the petroleum industry is inadequate.

The facts, as outlined above, clearly support the position that concentration at any level of petroleum operations is low, and that there has been rapid growth of the so-called "independents."

Changed Circumstances in the Industry Have Removed the Basis For Past Criticisms

Many of the factors, which led earlier critics such as the Federal Trade Commission to criticize the competitiveness of the industry, no longer exist. Allegations that the "major" oil companies had the power to control the price of crude oil have been superseded by the dominance of the Organization of Petroleum Exporting Countries (OPEC), which has gained complete control over international oil prices. Likewise, allegations that the "majors" had been able to prevent the inflow of foreign oil to the United States have been made irrelevant by the elimination of the Mandatory Oil Import Control Program. It had been alleged that the "major" oil companies had an incentive, through the depletion allowance, to shift profits from other operating levels into crude production, resulting in a squeeze on the "independent" refiners. Effective January 1, 1975, the depletion allowance, which in reality served as an economic incentive for the discovery and development of petroleum reserves, was eliminated for all but the smallest, non-integrated domestic oil producers. Even the Federal Trade Commission Staff acknowledges that this squeeze hypothesis "no longer exists." Also, it has been alleged that oil companies had taken advantage of state prorationing to keep crude oil off the market. But this allegation is a fiction since state production "allowables" have been at maximum levels for many years. In sum, important changes in the structure of the industry in recent years completely undercut the old arguments that the industry is noncompetitive.

Effects of Divestiture

In these times when huge capital investments are needed to insure the country's future energy requirements, divestiture would have an adverse impact on the borrowing capabilities of the industry. The non-integrated companies based entirely in one area of the industry resulting from divestiture would instantly have their risks increased. Many investments that were made when a company was integrated would be too risky for the fragmented companies to undertake when operations are confined by law to only one area of the industry and cannot be spread out over other integrated investments.

The borrowing abilities of these companies would be eroded because of the higher risks they would be assuming. For example, it is doubtful that funds previously available for an integrated refiner's capacity expansion program would be available at similar costs to a large refiner having no guarantees of crude supply or product marketability. Divestiture would raise the capital and operating costs for the vast majority of firms in the industry.

Millions of stockholders would be adversely affected by divestiture. A study sponsored by the American Petroleum Institute entitled, "Shareholders of the Six Largest U.S. Oil Companies" showed that 46 percent of the stockholders in these companies were retired. The firms were 90 percent owned, directly and indirectly, by 14 million people. The median estimated value of all common stock holding was $25,000; 44.7 percent of the direct shareholders had annual incomes of under $15,000, with the median income of all direct shareholders at $16,418; and 91 colleges and universities were stockholders. Divestiture would cause the value of the investments of these and other stockholders to fall. The companies that would be split off from the integrated companies would be economically less viable and there would be an increase in the risk element in the portfolios held by these stockholders.

The costs and difficulties of mechanically accomplishing divestiture would also add to the inadvisability of divestiture. The integrated company acts as one unit since its operations are melded together. Thus, for all practical purposes, in the event of divestiture, it would be virtually impossible to fairly allocate the assets and debts to each of the new firms created.

Divestiture would also affect the relationship of American companies with foreign sources of crude oil and create greater uncertainty at a time when stability of supply is essential. The cartel power of OPEC would be substantially strengthened by breaking up the private companies which are now large purchasers of their oil. Foreign competitors would still be able to benefit from the economies of integration and scale, giving them a competitive advantage over their U.S. counterparts.

Divestiture would make it impossible to achieve the goals of Project Independence. By increasing the risks and decreasing the efficiency of the companies involved in exploration, divestiture would curtail the flow of investment into this area and thus diminish rather than enhance the supply of crude oil and natural gas.

The integrated nature of the oil industry has helped the nation to maintain vital petroleum supplies in times of emergency because of its ability to respond rapidly as a result of its coordination at all levels of operation. However, if the industry were fragmented into many small non-integrated units, this efficient coordination would be eliminated, and it would be much more difficult for government to mobilize industry response.

There is no relationship between the problems presented by the current attempt at divestiture and the divestiture ordered in the 1911 Standard Oil Case.[1] The Standard Oil divestiture did not fragment operating companies. In Standard Oil, the courts ordered that the Standard holding company be dissolved, with its controlling shares in 33 geographically dispersed operating subsidiaries to be distributed on a pro-rata basis to the stockholders of Standard Oil of New Jersey. The geographical dispersion of these subsidiaries did not mean that the resulting companies were not vertically integrated. For example, Standard Oil of California continued to be a fully integrated company after the dissolution. It must be remembered that the Standard Oil Trust was a holding company, allowing its subsidiaries to maintain their own identities. This is not the case with today's integrated oil companies. These companies function as integrated units and do not maintain the operational integrity of their functional units.

* * * * * *

In these times when America needs to expand all of its productive facilities, policies that would intentionally disrupt the capital and exploratory expenditures of an industry as important as petroleum would prove disastrous. The U.S. economy needs to grow. This cannot be achieved through negative actions, but requires the establishment of a positive atmosphere in which innovation and expansion can be carried forward with some confidence.

The American petroleum industry through intense competition at every level has been and is now serving the American consumer extremely well.

---

[1]For an historical perspective of the Standard Oil breakup of 1911, see Chapter 5 by Hastings Wyman, Jr.

Proponents of a radical restructuring of the U.S. petroleum industry cannot carry the heavy burden of demonstrating that their alternative model would provide more efficiencies and productivity than the vertically integrated structure which exists today.

4

An Economist's Analysis

The Divestiture of Vertically Integrated Oil Companies

By Michael E. Canes

Deputy Director of Policy Analysis, American Petroleum Institute

Summary: Economic analysis casts doubt on the contention that vertical integration is a source of monopoly power and implies that firms facing monopoly power at some unintegrated stage of production can decrease that power through integrating into the stage, thus reducing retail product prices. Further, statistical evidence concerning the U.S. petroleum industry tends to refute charges that vertical integration there has resulted in monopoly.

Economies of vertical integration stem from the particular nature of petroleum related investment which is specialized and immobile and requires an assured flow of oil. It also stems from diversification of risk resulting in lower capital costs and from saving in transaction costs of buying and selling the intermediate products.

A number of bills recently have been introduced in the U.S. Congress which if enacted would compel large integrated oil companies to divest themselves of parts of their operations. For example, Senate bill 745 would modify the Interstate Commerce Act to compel refiners owning over 25 percent of their crude oil inputs or selling over half their product outputs through their own distribution channels to divest their production, marketing and transportation assets. Senate Bill 756 similarly would amend the Clayton Act. Senate bills 1137 and 1138 would compel integrated companies to divest their marketing and pipeline assets, and Senate bill 2387 would compel separation of production, refining, transportation and marketing operations by "major" integrated oil companies.[1] Similar bills are pending in the House of Representatives.[2]

The stated purpose of these bills is to enhance competition in the domestic oil industry by reducing the extent of vertical integration there. The specific question raised by the bills is whether the social benefits of oil company vertical integration outweigh the costs or vice versa. And since vertical integration is a widespread phenomenon among U.S. companies, the broader question is whether the phenomenon should be generally discouraged under force of law. This chapter addresses the costs and gains of oil industry vertical integration.[3] Many economists have argued that vertical integration does not decrease competition and in fact works to reduce misallocative effects of monopoly. However, others have argued that vertical integration is a strategy employed to discourage entry or encourage exit from an industry. Implications of this assertion are compared to evidence from the U.S. oil industry. According to this evidence, vertical integration has not had anticompetitive consequences for that industry.

The phenomenon of vertical integration then is explained in order that sources of economies can be made explicit. It is pointed out that while many U.S. oil firms integrate some operations, these companies also transact with independent firms, and that this pattern of both integration and market transacting is consistent with the realization of cost savings.

Some special short-term problems concerning competition between integrated and non-integrated U.S. oil firms then are discussed. These problems are seen to be the result of changed market conditions of crude oil and certain past federal support policies, but are not the result of vertical integration within the industry. The policy dilemma is whether to continue the past support policies, not whether to compel vertical divestiture.

The major finding of the chapter is that there are no discernible consumer benefits from oil industry divestiture whereas costs are certain to be large. Since a relevant criterion for social policy is whether consumers are made

---

[1]In S. 2387, major oil companies are defined as those which produced 36,500,000 barrels of crude, condensate, and natural gas liquids or 200 billion cubic feet of natural gas in 1974, *or* which refined 75 million barrels of product within the U.S. *or* which distributed 110 million barrels of product within the U.S. As formulated, the bill would affect the twenty largest U.S. oil companies.

[2]See for example H.R. 509, 310, and 5213, all of 1975-1976.

[3]Vertical integration in the oil industry refers to the simultaneous operation of crude oil production, refining, crude and product transportation, and product marketing by a single firm. There are around thirty large U.S. companies engaged in all four of these businesses and many more which engage in at least two. In addition, large numbers of independent firms operate in each phase of the oil business.

better off, the major implication is that no case can be made for the vertical divestiture of U.S. oil firms.

Competitive Aspects of Oil Industry Vertical Integration—An Economist's Views

A divestiture policy would provide consumer benefits if it resulted in reduced prices for oil products. Prices would be reduced if oil company vertical integration were a source of monopoly power, since divestiture then would reduce that power. Reduced market power is the stated purpose of the various divestiture bills now before the U.S. Congress.

The economics literature casts doubt on the contention that vertical integration is a source of monopoly power. Instead, several economists have argued that were monopoly to exist at one stage within an industry the vertical integration of the monopolized stage with other stages would benefit consumers.[4] In essence, they argue that firms facing monopoly power at some unintegrated stage of production can reduce that power through integrating into the stage, or that integration of a monopolized and competitive stage induces more efficient production techniques. In either case, the result of integration is to reduce retail product prices.

Economists also have argued that were monopoly to exist at two or more stages of production, the vertical integration of those stages would benefit consumers.[5] In this case, consumer prices are reduced because the integrated firm would choose to exert monopoly power at but one stage whereas successive monopolists would do so at each stage.

Finally, economists have argued that where competition exists at each of two or more stages of production, the vertical integration of those stages will not achieve monopoly power.[6] The reason is that under competitive conditions, a nonintegrated firm will have substantial choices among sup-

---

[4]See Burstein, M.L., "A Theory of Full-Line Forcing," *Northwestern University Law Review*, 55 (Feb. 1960), 62-95; Stigler, George J., "The Division of Labor is Limited by the Extent of the Market," *Journal of Political Economy*, Vol. 59 (1951), pp. 185-193; Mead, David E., "Vertical Integration, Monopoly, and Fixed Coefficients," unpublished manuscript, Federal Energy Administration, Sept, 1975; and Vernon J.M. and Graham, D.A., "Profitability of Monopolization by Vertical Integration," *Journal of Political Economy*, (July/August, 1971); 924-925.

[5]See Spengler, J.J., "Vertical Integration and Antitrust Policy," *Journal of Political Economy*, 58, 1950, 347-358; Liebeler, Wesley J., "Towards a Consumer's Antitrust Law: The Federal Trade Commission and Vertical Mergers in the Cement Industry," *UCLA Law Review*, Vol. 15, No. 4, June 1968.

[6]See, for example, Peltzman, Sam. "Issues in Vertical Integration Policy," in *Public Policy Toward Mergers*, Sam Peltzman and J. Fred Weston (eds.), Pacific Palisades, California, 1969, 167-176.

pliers and customers and so cannot be foreclosed from operating by an integrated rival. The policy implication of this argument is that if competition exists at the various stages of a good's production, the vertical integration of the stages is not socially harmful.

A substantial body of evidence has been accumulated to test whether the various stages of the U.S. oil industry are competitive. Such evidence can be categorized by whether it pertains to industry structure, ease of entry, or rate of profit.

By the usual structural tests, all phases of the industry are competitive. Evidence presented in Chapter 2 (see especially exhibit 3 and accompanying text) gives strong indication that economic power is not concentrated in the domestic petroleum industry.

Entry into the four industry stages appears to be relatively unrestricted. State regulation of crude oil output has not prevented new firms from entering the producing sector of the industry. Rather, the number of oil producers has tended to vary with the real price of crude, rising during the 1950's while crude prices increased or held steady and falling during the 1960's and early 1970's as the real price of crude declined.

There are certain restrictions on entry into refining, but these mostly have to do with government site and environmental regulation. Over the past few years, at least thirteen refiners have been delayed or turned down in attempts to locate new refineries along the U.S. East Coast. Nevertheless, since 1951, thirteen companies have entered the business with 50,000 barrels/day capacity or more and nine others which had less than 50,000 barrels/day capacity in 1951 have added 50,000 barrels/day capacity or more.[7]

Although interstate oil pipelines are regulated as common carriers, there are no legal restrictions against entry, and both oil and non-oil firms have entered this market over the past twenty years. Between 1951 and 1972, the number of firms owning interstate pipelines increased from 65 to about 90. Further, five of the top 20 firms of 1972 were not among the top 20 firms in 1951, and have since captured at least 1% of the market. Entry into the world oil tanker market is similarly unrestricted, and the larger tanker firms have steadily lost market share to entrants and smaller firms over the past two decades.[8]

---

[7]See Exhibit D, Chapter 7, page 112. In 1951, the U.S. Department of the Interior classified twenty refiners with 50,000 barrels/day or more of capacity as "majors." By this definition, an additional twenty-three companies became major refiners between 1951 and 1974.

[8]See Neil H. Jacoby, *Multinational Oil,* Macmillan Publishing Co., 1974, table 9.10, p. 202. According to data in that table, six of the top 20 firms (by tonnage) in 1972 were not among the top 20 firms in 1953.

Entry into wholesale and retail operations is particularly unrestricted because capital requirements are lower than in other phases of the oil business. In recent years the marketing sector of the oil industry has been subject to declining profitability and because of this the total rate of entry has declined. Nevertheless, over an 18-state sample, an average of about four firms per state entered the retail gasoline market between 1968 and 1974.[9]

The profitability of the oil business appears to be about the same as that of other businesses (see Exhibit 1). Presumably, if there were monopoly power within the industry, such power would be manifest in higher than normal rates of profit.

The relative profitability of the oil industry has been estimated via use of stockholder rate of return data and via accounting returns on equity. Professor Edward J. Mitchell compared shareholder rates of return for domestic refiners, domestic producers, and internationals with those for Standard & Poor's 500 Stock Composite over the periods 1953-1974 and 1960-1974.[10] His results show *lower* than average returns for all of the various oil company categories over the 1953-1974 period; and lower than average returns for the producers and internationals but higher than average for the domestic refiners over the 1960-1974 period. Taking the companies as a whole, their shareholder returns were below average for 1953-1974 and about average for 1960-1974.

First National City Bank annually publishes ten-year average accounting returns on equity for a large number of industries. For the years 1965-1974, this source reported a 13.4 percent average return for the petroleum industry, slightly above the average for all manufacturing (13%) but below the average return on mining (14.7%). For the periods 1964-1973 and 1963-1972, the Bank reported lower returns for the petroleum industry than for manufacturing generally and about the same as for mining. A comparison contained in the Ford Foundation Energy Policy Project showed that over the shorter period 1967-1972, the average accounting return on equity to the largest 20 oil firms (10.8%) exactly

---

[9]See Michael E. Canes, "Vertical Integration, Competition and Monopoly," unpublished paper, American Petroleum Institute, November 1974. Entry is defined in the paper as having at least 1% of the state market in 1974 *and* less than 1% in 1968.

[10]See Statement of Professor Edward J. Mitchell, submitted to the Subcommittee on Antitrust and Monopoly, Committee on the Judiciary, U.S. Senate, August 6, 1974. Mitchell classified most integrated domestic companies as refiners.

equalled that to all manufacturing. None of these numbers is consistent with monopoly returns being entered by the oil industry.[11]

Despite both theory and evidence contrary to the assertion that oil industry vertical integration is associated with monopoly power, the belief persists that such as association exists. Specifically, it has been asserted that integration of crude oil production and pipeline transportation with refining and marketing has enabled large existing oil firms to foreclose new refiners and marketers from entering the oil business while encouraging existing competitors to exit.[12] This foreclosure allegedly has been accomplished via refusal to sell or transport crude oil or products to independent refiners or marketers. If this assertion were true, then the market share of the large vertically integrated refiners and marketers should have been stable or increasing over time. However, existing evidence indicates that the opposite has occurred. From 1951 to 1974 the market share of the largest twenty U.S. refiners in 1951 fell from about 81% to 71%.[13] Reliable marketing data are available for a lesser period. According to these data, the market share of non-major company affiliated retail gasoline marketers grew from about 20% in 1968 to about 28% in early 1975. Further, the integrated companies have lost market share in refining and marketing while their share of U.S. crude production has grown (see Exhibit 3). This evidence is inconsistent with the thesis that major refiner integration into crude production forecloses independent refiners and marketers from operating profitability.

Both economic theory and evidence refute the contention that oil industry vertical integration is associated with monopoly power. In consequence, vertical divestiture could not be expected to reduce prices to oil consumers. From this reasoning, there are no discernible consumer benefits from oil industry divestiture.

Economics of Oil Industry Vertical Integration

Every commodity passes through multiple stages of production in its transformation from raw material to finished product. The meaning of ver-

[11]See Duchesneau, Thomas D., *Competition in the U.S. Energy Industry,* A Report to the Energy Policy Project of the Ford Foundation, Ballinger, Cambridge, Mass., 1975, p. 157.

[12]See the Preliminary Federal Trade Commission Staff Report on its Investigation of the Petroleum Industry, reprinted as "Investigation of the Petroleum Industry" by the Permanent Subcommittee on Investigations of the Senate Committee on Government Operations. 93rd Congress, 1st Session, July 12, 1973. For a point by point rebutal to the FTC staff report, see Staff Analysis of Office of the Energy Advisor, Dept. of the Treasury, August 27, 1973.

[13]See Exhibit 2.

tical integration is that more than one of these stages of production is conducted by a single firm. A production line exemplifies such integration, each worker along the line carrying out some specialized operation before turning the commodity over to a subsequent operation. Vertical integration occurs because such a production line generally is operated by a single firm employing many workers and not by many firms, each owning a small portion of the line and conducting one specialized phase of the production process.

The production-line example can be used to illustrate the nature of cost savings from vertical integration. Suppose instead that the vertical integration of production were illegal. Because of this, many separate firms would carry out the same production process formerly carried out by the single firm. Each of these separate firms would buy the good in some intermediate stage, conduct a specialized operation, and then sell the good to another in a slightly more advanced stage.

In contrast to the previous situation, the profitability of each stage now would depend not only on efficiency in production but also on efficiency in buying and selling the good at its intermediate stages. Each firm can benefit from buying at lower prices and selling at higher, and therefore each would expend resources to obtain skills useful to such purpose.

The types of skills involved include those of purchasing agents, contract negotiators, lawyers, salesmen, and accountants, and the firms taken together will employ more of such skills than did the previous single firm. Thus, not only are ownership arrangements shuffled but the physical nature of production is changed. The result of carrying out production in this way is to increase the number of transactions involving the good, the total cost of transacting for it, and hence the total cost of producing it.

Costs of production will increase further if technologies which are economically attractive with vertical integration no longer are so when separate firms conduct different production phases. It would be possible for separate firms jointly to invest in and own a single production line. But joint firm investment is economic only where gains from risk dispersion outweigh costs of arranging common investment and operating policies. Even if it were economic for a single firm to invest in the production line, multifirm arrangements could be less attractive than the use of a completely different technology, e.g. one in which firms use physically separate operating facilities.

Although the vertical integration of production yields costs savings, few goods are produced under totally integrated conditions. For example, producers of crude oil generally do not themselves produce drilling or seismic equipment and oil refiners usually do not build refineries and often pay others to maintain and repair them. Also, few refiners are totally self-sufficient in crude oil or own sufficient marketing outlets to absorb their entire refinery output.

There are at least three reasons why firms do not fully integrate the production of a good. First, there are both gains and costs of larger firm size. Gains from larger size generally outweigh costs over a range of production rates, but eventually costs outweigh gains. Since a firm that integrates into more than one stage of production generally will be larger than if it engages in but one, the gains from integration eventually will fall relative to costs as firm size increases.[14]

Second, there are diseconomies of engaging in multiple businesses. These diseconomies arise because a vertically integrated firm's management must gain knowledge of multiple production technologies and of conditions in multiple input and output markets. For this reason, integration is more economic the more closely related are successive production stage technologies and the smaller the increase in numbers of input and output markets.

And third, if market demand or supply conditions fluctuate with uncertainty, a completely integrated firm has reduced flexibility to employ more or less inputs in each of its component units as its market opportunities change. One economic function of "independent" firms in such markets is to supply more services to integrated firms whose demand has risen and less to those whose demand has fallen, and over time such an independent will deal with several different integrated firms.

Economies of vertical integration in the oil industry stem in part from the particular nature of investment and profitability in that industry. Very large amounts of investment capital are required to take advantage of scale economies in refining and pipeline transportation, but once built refineries and pipelines are specialized in purpose, durable, and virtually immobile. These characteristics subject investment in refining and pipeline transportation to the risk that very low returns would be earned were flows of crude oil or oil products to be interrupted or reduced. At some price in terms of additional oil buyers, product salesmen, contract lawyers, inventories, etc., a refiner can reduce the likelihood of costly oil-flow interruptions, but generally it is cheaper for him to control at least some crude oil sources, transportation means, and product outlets. Similarly, an oil pipeline owner can reduce the risk of reduced oil flows through means other than integration, but generally the costs of doing so make it economic for him to control

[14]There appear to be economies of scale in the oil industry which specifically result from vertical integration. This point is taken up below with regard to refinery and pipeline investment.

at least some of the sources and outlets for the pipeline's oil flow. Students of oil industry vertical integration long have argued that "assurance" of a flow of crude oil and product is fundamental to refinery or pipeline economics. The argument here is that such assurance is expensive to obtain, and that under fairly general conditions vertical integration is the most economical method to do so.

A second economy arises from the relationship between the profitability of crude oil production and those of refining, marketing and transportation. Studies of the oil industry have shown that profits in these various industry operations are imperfectly correlated so that integration of these functions tends to reduce the variability of an oil firm's profits.[15] Because of this, integration provides a real service to investors, that of diversification of risk, and for this reason the cost of capital to an integrated oil firm is reduced. Such a reduction allows greater investment in all phases of oil industry operation than otherwise would be the case.[16]

The above described benefits of vertical integration also pertain to investment in other kinds of refinery and transportation inputs and to further integration with crude oil sources and marketing outlets than actually occurs. For example, there would be less need to shop for steel storage tanks, steel pipelines, etc. if oil companies manufactured these themselves. But markets for fabricated steel products are well developed so that the gains from integration are small relative to the costs of involving oil company management in areas removed from their specialized expertise.

Similarly, complete integration of production, refining, transportation, and marketing would reduce costly reliance on market sources, but is not economic because of reduced flexibility to vary the employment of resources in each of these stages with unanticipated changes in demand and supply. One important function of independent firms in the oil industry is to supply services on an "as needed" basis, where these needs vary among integrated companies over time.[17] For example, integrated oil firms own

[15]See McLean, J. and Haigh, R. *The Growth of Integrated Oil Companies,* The Harvard University Press, 1954.

[16]Of course, investors can themselves diversify risk through holding shares of oil companies separately engaged in different phases of the business. But given that integration has other advantages, company risk diversification is a further real service. Also, by earlier arguments, company integration reduces business risk and increases absolute company size. Both tend to reduce company costs of capital.

[17]There are other reasons for the existence of independent oil firms. Certain oil products lend themselves to highly specialized processing operations, and there are advantages to specialization both in woldcat drilling and in the retail marketing of oil products. Most branded and non-branded gasoline dealerships are independently operated because efficient marketing at the retail level involves both detailed knowledge of local demand conditions and strong incentive to act in accordance with this knowledge.

some oil tankers but contract for others, and over a period of years the same independently owned tanker is used by several integrated firms. Because a combination of vertical integration and contracts between integrated and nonintegrated firms allows both to utilize their resources more efficiently, an industry which contains both types of firm structure reduces costs of supplying oil products to consumers.

The forced divestiture of integrated oil operations would not bring the industry to a halt. But as with the simple production-line example above, such divestiture would increase costs of engaging in the oil business. Contracts and other buyer-seller arrangements would replace integrated companies, and the non-integrated units would seek to profit through the gaining of bargaining advantage and the employment of negotiating skill, purchasing and selling expertise, etc. More resources would be expended to gain information about buying and selling opportunities and about alternatives available to suppliers or customers and more would be spent in arranging, implementing and enforcing buyer-seller contracts. The total cost of transacting would rise in the oil industry, and since the cost of transacting is a cost of production, so too would the cost of supplying oil products.

A second consequence is that the cost of capital to oil firms would increase. Divestiture would increase the business risk of operating at each stage of the industry, would reduce the equity base available to back any given investment program, and would foreclose opportunities for oil companies to diversify their investments among industry stages. To the extent that these factors increase capital costs, there would be reduced investment in each of the separated parts of the industry. Further, it is plausible to assume that technologies now in use in the industry would not survive. In particular, because of increased uncertainty concerning crude supplies and market outlets, investors would be less willing to risk money in large-scale refinery or pipeline projects, so that opportunities to realize scale economies in these markets would be reduced.

Another consequence of divestiture is that unintegrated U.S. oil companies would have to compete in international markets with fully integrated foreign firms. Not only are there privately held such companies (e.g. Royal Dutch Shell), but foreign governments have almost uniformly opted to organize their state oil companies as fully integrated concerns.[18] Because there are real economies associated with vertical integration, U.S. companies would tend to do less well than their integrated competitors in ob-

[18]Examples are PEMEX, ENI, and ELF ERAP. The Norwegian government intends that Statoil, its newly set up state oil corporation, become fully integrated, and both the British and Swedish governments have announced similar plans for their proposed state oil companies. Several Arab governments also have announced plans to build fully integrated organizations.

taining drilling concessions, setting up and operating foreign refining and marketing operations, etc.

Finally, it is unclear whether vertically non-integrated U.S. firms would respond as efficiently as integrated to a renewed boycott of the U.S. by Arab oil producers. Non-integrated firms rely more on short-term price signals to determine their courses of action, but recent experience indicates such short-term signals can be misleading under boycott conditions.

The above consequences assume an equilibrium after divestiture has occurred. But there also would be costs associated with a transition from the present structure of the industry to a new one. These costs would include those of carrying out individual company divestiture plans (e.g. of categorizing and separating assets, of selling or otherwise distributing them, of setting up new operating managements, etc.), of retiring current debt and equity instruments, and of raising capital in the face of extraordinary investor uncertainty. Whereas existing integrated companies have given investors substantial evidence of their economic viability, newly separated units would require several years to establish track records. As a result, it is unlikely that any part of a now integrated firm's operations would expand as rapidly during and soon after a period of divestiture as without such action.

Ultimately, higher costs of producing, processing, transporting and selling oil would be borne both by U.S. consumers of oil products and by owners of oil companies. If divestiture negates economies derived from vertical integration and there are no compensating reductions in cost, then prices for oil products would rise. If integrated foreign oil companies could freely compete in the United States after divestiture by domestic firms, then a greater share of the market would accrue to the foreign firms and prices would rise only to the extent integrated U.S. firms could have operated more cheaply. However, if foreign oil firms were precluded from operating as integrated entities in the U.S., then U.S. firms would maintain market share but consumers would face still higher prices for oil products.

Short Term Competition Between Integrated and Independent Oil Firms

Over the past several months, representatives of some independent refining and marketing companies have testified before the Congress that an end to crude oil price controls and allocation would harm these independent sectors and consequently would reduce competition in the industry. The essence of their argument has been that decontrol will reduce access to low cost crude oil for the independent sector and would increase this access for large vertically integrated companies. Several of the representatives concluded that divestiture would serve to equalize crude oil access and thus would enhance competition in the oil industry.

Despite these arguments, the issue of divestiture is very different from that of competition between integrated and non-integrated refiners and marketers after removal of crude oil price and allocation controls. There is indeed a social policy issue concerning competition under such changed circumstances, but it pertains to whether smaller refiners should continue to receive differential government support given that they and the marketers they serve have made investments on the basis of such support in the past. This issue deserves serious consideration, but it is not the same issue as whether vertically integrated oil companies should be forced to divest.

The ending of price and allocation controls over crude oil implies that all domestic refiners will pay open market prices for such oil. Since the producing units of integrated companies would have the option of selling on the open market, there is no reason to expect them to transfer oil at lower than market prices to their own refineries. Rather, under open market conditions, refiners and marketers who most efficiently carry out their operations will best survive, and there is no necessary reason why these would be only integrated refiners and marketers. Evidence from the period before price controls indicates that non-integrated companies are capable of expanding their market shares under such conditions.

On the other hand, the ending of price and allocation controls would end differential government support for independent refiners that began within the Mandatory Oil Import Quota Program in 1959 and has continued through the Federal Energy Administration's crude oil entitlements program in 1975. Under the oil import quota program, smaller refiners were given tickets to import oil in disproportionate numbers to their refining capacity.[19] Since these tickets had value, this amounted to government support for the independent refining sector. Although the quota program ended in 1973, support for independent refiners has continued via legislated access to price-controlled domestic crude oil. The present method of support is through the entitlements program, which guarantees all refiners access to the national proportion of price-controlled oil to total oil refined. Through this program, refiners with below average access to price-controlled crude oil are allocated rights which they in turn sell to refiners with above average access to such oil. Large integrated refiners tend to have above average access and independent refiners below average.

[19]The exact percentage going to different sized refiners varied from year to year; see *The Oil Import Question,* Report on the Relationship of Oil Imports to the National Security by the Cabinet Task Force on Oil Import Control, February 1970, p. 12.

The distinction between access to government support and to crude oil under open market conditions is borne out by a recent Federal Trade Commission survey of U.S. refiners.[20] According to the survey, the great majority of refiners favor decontrol of domestic crude oil and almost none fear loss of access to crude oil. These results were obtained even though the eight companies currently involved in litigation with the FTC were not included in the survey.[21] However, according to the FTC, many independent refiners favor gradual decontrol (which would retain government support over a longer period) and many feel that with the end of controls some other form of government support for the independent refining sector should be instituted.

The issue concerning support is whether it should be continued in the light of past investments made on the basis of then existing support programs. The argument can be made that abrupt changes in present support policies weaken incentives to respond to future policies. This policy problem deserves careful attention.[22] But the issues involved are different from those of vertical divestiture and should be treated separately.

Findings and Implications

The U.S. Congress currently is considering legislation which would compel vertically integrated oil companies to divest themselves into separate producing, refining, marketing and transportation companies. The immediate question is whether the social gains from such legislation outweigh the social costs. The broader question is whether vertical integration should be tolerated in the United States as an industrial organizational structure.

The argument that oil company vertical integration enhances monopoly power in the refining and marketing industries was examined. Economic theory does not support the argument, and it was pointed out that an important implication—that major integrated firms will hold or increase their market share—is contradicted by available empirical evidence. Because both theory and evidence concerning vertical integration are inconsistent

---

[20]FTC Staff Report on the Effects of Decontrol on Competition in the Petroleum Industry, September 5, 1975.

[21]These companies are Exxon, Gulf, Mobil, Standard Oil of Indiana, Shell, Standard Oil of California, Atlantic Richfield, and Texaco. Most of these companies would benefit from decontrol and presumably would have supported such a policy.

[22]A possible solution has been suggested by the Administration. According to the *Oil and Gas Journal* (September 15, 1975, p. 84), in the event no phaseout program of crude oil price controls is enacted, the President will introduce legislation to continue subsidies to small refiners but will phase these out over three years.

with assertions of monopoly power, vertical divestiture cannot reduce such power and hence such policy cannot make consumers better off.

It was then argued that an oil industry which includes both vertically integrated and non-integrated companies will minimize costs of producing, processing, transporting and selling oil in the United States. Economies derived from oil company vertical integration were described, but it was pointed out that independent oil companies also serve to reduce total costs of supplying oil to consumers. By this reasoning, costs are minimized by a mix of company structures, and forced divestiture would raise total costs of oil supply in the U.S. Ultimately, such a cost increase would be borne by oil company shareholders in the form of reduced profits and by consumers in the form of higher prices for oil products.

Finally, the issue of oil industry competition after an end to crude oil price and allocation controls was briefly examined. Despite claims by some independent refiners and marketers that divestiture would enhance such competition, the issue is more properly viewed as whether past government support policies for the independent sector should be continued, at least over the short term. This issue is separate from the divestiture issue.

The above findings imply that oil industry vertical divestiture would be an erroneous social policy. Presumably the overall policy aim is to encourage an industry structure that will supply oil to consumers at minimum cost. Forced divestiture would raise this cost with no discernible consumer benefits. On the other hand, retention of the present diversity of oil company structures would be consistent with the policy aim.

EXHIBIT 1

PROFIT STUDIES OF THE DOMESTIC OIL INDUSTRY

	1953-74	1960-74
Shareholders Annual Average Rate of Return[1]		
Standard & Poor's 500 Stock Composite	12.4%	8.3%
Domestic Refiners	10.8%	10.4%
Domestic Producers	6.4%	4.5%
Internationals	10.2%	7.0%
All Oil Companies	10.2%	8.3%

	1963-1972[2]	1964-1973[2]	1965-1974[2]
Accounting Return on Equity			
Petroleum Production & Refining	11.8%	12.3%	13.4%
Total Mining	11.7%	12.5%	14.7%
Total Manufacturing	12.2%	12.6%	13.0%

	1967-1972[3]
Twenty Largest Petroleum Companies	10.8%
All Manufacturing (1967-1971)	10.8%

[1]Source: Statement of Professor Edward J. Mitchell. Submitted to the Subcommittee on Antitrust and Monopoly, Senate Committee on the Judiciary, August 6, 1974.

[2]Source: First National City Bank, Monthly Letter, April of each year.

[3]Source: Thomas D. Duchesneau, *Competition in the U.S. Energy Industry*, A Report to the Energy Policy Project of the Ford Foundation, Ballinger, Cambridge, 1975, Table 4-3, p. 157.

EXHIBIT 2

1951 AND 1974 MARKET SHARES OF TWENTY LARGEST REFINERS OF 1951 (by refining capacity)

	January 1, 1951[1]		January 1, 1975[2]	
	Total U.S. Capacity	6,963,644		15,770,000
1.	Standard (N.J.)	769,500	Exxon	1 243,000
2.	Socony Vacuum	548,500	Mobil	903,500
3.	The Texas Co.	510,000	Texaco	1,073,000
4.	Standard (Indiana)	502,500	Standard (Ind.)	1,115,000
5.	Gulf	458,500	Gulf	904,200
6.	Standard (California)	388,200	Standard (Cal.)	952,000
7.	Shell	382,000	Shell	1,150,000
8.	Sinclair[3]	360,000		
9.	Cities Service	215,000	Cities Service	268,000
10.	Sun	200,000	Sun	569,000
11.	Tide Water	192,500	Getty	218,731
12.	Phillips	182,500	Phillips	408,000
13.	Atlantic[4]	167,000	Atlantic Richfield	724,600
14.	Union	153,400	Union	496,000
15.	Standard (Ohio)	116,500	Standard (Ohio)	431,000
16.	Pure[5]	113,300		
17.	Richfield[4]	110,000		
18.	Ashland	105,700	Ashland	362,943
19.	Continental	93,150	Continental	377,000
20.	Texas City	50,000	Texas City	74,500
	Total	5,618,250		11,249,794
	Percent of U.S. Total	80.68%		71.47%

[1]Source: Statement of Donald C. O'Hara, President, National Petroleum Refiners Association, before the Subcommittee on Antitrust and Monopoly of the Senate Committee on the Judiciary, August 8, 1974.

[2]Source: Statement of Donald C. O'Hara, President, National Petroleum Refiners Association, before the Subcommittee on Antitrust and Monopoly of the U.S. Senate Judiciary Committee, November 12, 1975.

[3]Merged with Atlantic Richfield during the period.

[4]Atlantic and Richfield merged during the period.

[5]Merged with Union during the period.

EXHIBIT 3

U.S. MARKET SHARES OF TOP 4, TOP 8, AND TOP 20 REFINERS OF 1951[1]

	Refining Capacity[2]				Gasoline Marketing[3]		
	Top 4	Top 8	Top 20		Top 4	Top 8	Top 20
1951	33.5	60.2*	80.7	1969	30.1	56.7	76.3
1974	27.2	51.1	71.5	1974	29.0	51.8	69.7

Net Crude and Condensate Production[4]

	Top 4	Top 8	Top 20
1950	18.2	33.8*	45.8
1974	23.9	42.2	56.7

*Including Atlantic, Richfield, and Sinclair as one company.

[1]Top 4 refers in all cases to Exxon, Mobil, Texaco and Standard Oil of Indiana. Top 8 also includes Gulf, Standard Oil of California, Shell and Sinclair-Atlantic-Richfield (Sinclair was the 8th largest U.S. refiner in 1951. Sinclair, Atlantic, and Richfield since merged into a single company).

[2]Source: Statements of Donald C. O'Hara, President, National Petroleum Refiners Association, *op. cit.*

[3]Source: *National Petroleum News,* Mid-May issues for 1971, 1975.

[4]Source: 1950 company figures from Moody's Industrial Manual; 1974 company figures from company annual reports, U.S. totals from U.S. Bureau of Mines, *Petroleum Statement.*

5

The Standard Oil Breakup of 1911 and Its Relevance Today

By Hastings Wyman, Jr.

Attorney-Writer, American Petroleum Institute

Summary: The economic conditions which led up to the breakup of the Standard Oil Company of New Jersey in 1911 were entirely different from those prevailing today. Whereas Standard Oil dominated virtually the entire oil industry in its time, today the concentration ratios of neither the top 4 nor the top 8 companies can match Standard's earlier dominance. In addition, the present 14 million shareowners of the top six companies contrast sharply with 16 individuals owning 80 percent of Standard Oil, or only 6,000 shareholders in total. Another contrast is Standard's high profit rates compared to normal profit rates today.

This chapter compares conditions in the oil industry leading up to and immediately following the 1911 breakup of the Standard Oil Company with conditions prevailing today to see if the earlier action is a valid precedent for contemporary policy. The early history of the petroleum industry in general and the Standard Oil Company in particular is briefly discussed in order to put the 1911 divestiture into historical perspective.

The dominance of the petroleum industry by one company, the Standard Oil Company, which was the reason for the divestiture movement early in this century, is not a reality today. Whether we are talking about concentration ratios or profit rates or patterns of ownership, the contrast in conditions between 1911 and today is marked. The similarities are few. It is also clear that the economic and political factors that encouraged the growth and prosperity of oil companies after 1911 are not reflected in this time of energy crisis. Because of this, the policy solutions of 1911 would not be appropriate today. Although nostalgia for less complicated times with relatively simple solutions to social and economic problems may be understandable, discussion of energy policy today should center on more appropriate actions to solve the nation's energy problems.

The Early History of the Oil Industry

While oil had been used for medicinal and miscellaneous purposes prior to the drilling of the first well by Colonel Drake in 1859, its uses were strictly limited because it could not be obtained in large quantities. But the uses of even limited amounts were well known, and in 1855 Professor Silliman at Yale had published a study showing the feasibility of refining illuminants, lubricants, gas and paraffins from petroleum.[1] After oil was first discovered in quantity, the growth of the oil industry was meteoric.

It has been said—accurately—that oil had just been discovered when Lincoln was nominated for President in 1860 and that virtually the entire civilized world depended on it by the time Lincoln was assassinated in 1865.

The industry's first few years were marked by successful solutions of one seemingly insurmountable obstacle after another in rapid succession. Once oil had been found and a method devised to get it out of the ground, the most immediate problem was transporting it to refineries.

One very early answer was the construction of pipelines. The early pipelines were often dug up and destroyed by teamsters who correctly perceived the effect on their own business. However, the first successful one was built in 1865 and was five miles long.[2]

These early pipelines carried the crude oil to the river port to be shipped East, to local refineries, or to railroad terminals.

Soon rail transport came to be relied on almost exclusively, with pipelines connecting the oil fields to the railroad heads. This development reduced the costs of transporting crude oil dramatically. And while rail transportation was later supplanted by long distance pipelines, the railroads played a key role in the development of the oil industry and how it was organized.[3]

The man who later came to dominate the oil industry and became a symbol for both the sins and the virtues of capitalism began his career in the wholesale produce business.

After an early business trip to the oil regions in 1862, John D. Rockefeller went into the refining business with several partners.

---

[1]Ida M. Tarbell, *The History of The Standard Oil Company* (New York: McClure, Phillips Co., 1904), Volume I, p. 7.

[2]Jerome T. Bentley, *The Effects of Standard Oil's Vertical Integration Into Transportation...*, (Unpublished thesis for Ph.D., University of Pittsburgh, 1974) p. 14.

[3]*Ibid.*, p. 13.

On January 10, 1870, John D. Rockefeller, his brother William, and three other partners formed the Standard Oil Company. The company was capitalized at $1,000,000, and John D. Rockefeller was elected the first President.[4]

The Rebate and the Standard Oil Monopoly

Rebates—or kickbacks of a percentage of the standard price to preferred customers—were a common business practice during this era. Many observers, especially Rockefeller's contemporary critics and competitors, believed that Rockefeller's negotiation of a lower transport fee was the chief method by which he came to dominate the oil industry. The railroads, chiefly the Erie and the New York Central, justified the practice by the economies of scale achieved when a shipper could guarantee a minimum shipment of large volume. Nevertheless, the practice clearly discriminated in favor of the large user who was thus able to bring his product to the market at significantly less cost than his competitors.

Whether it was the rebate, other factors, or some combination thereof, by the early '70s Rockefeller's Standard Oil owned virtually every refinery in Cleveland, which became the leading refining city.[5] And by the end of the decade, Pittsburgh, Philadelphia, New York and the oil regions came under Standard's domination as well.[6]

Construction of the Standard Oil Trust

Rockefeller and his able associates continued to buy other companies—either for cash or exchange for stock in Standard Oil.

In 1879 the stocks of the various companies held by Standard were transferred to a small group of trustees. This was the first time a trust agreement was used for such purposes[7] and gave the name "trust" its popular business usage.

In 1892, the Supreme Court of Ohio declared the trust illegal and for a period of about seven years, the various Standard companies were held together by little more than the common interest of the stockholders rather than a formalized structure. In 1899, the company reorganized under the

---

[4]Allan Nevins, *Study in Power: John D. Rockefeller, Industrialist and Philanthropist*, (New York: Charles Scribner's Sons, 1953), Volume I, p. 83.

[5]*Ibid.*, p. 159.

[6]*Ibid.*, p. 230.

[7]*Ibid.*, p. 385.

laws of the State of New Jersey and became the Standard Oil Company (N.J.).[8]

The Political Climate

The political climate in which Standard operated had been growing steadily more hostile since the decade of the '70s. The tremendous economic growth of the second half of the nineteenth century—capitalization per business establishment multiplied eight times between 1850 and 1880—created tensions that were reflected in the politics of the day. Henry Demarest Lloyd wrote *Wealth Against Commonwealth,* which was published in 1894. Ida Tarbell's famous *The History of The Standard Oil Company* was first serialized in *McClure's Magazine*, and published in book form in 1904. Theodore Roosevelt was making his reputation as a "trust-buster," and the Democrats were equally vehement in denouncing "these commercial monsters called trusts."

In 1890, the Sherman Antitrust Act was passed. Passage of the Clayton Act followed.

In 1906, the government filed suit under the Sherman Act, charging Standard with conspiring "to restrain the trade and commerce in petroleum, commonly called 'crude oil,' in refined oil, and in the other products of petroleum." Based on the Bureau of Corporations report on Standard Oil, the thrust of the charge was that Standard had established a monopoly in refining through its position in transportation.

In the meantime, various other trials involving Standard were going on simultaneously. Between 1904 and 1906, over 23 suits were filed by state governments against Standard.

Comparison With Today

The breakup of Standard Oil in 1911 is often cited as a precedent for requiring vertical disintegration of oil companies today. The fact is that the competitive structure of the oil industry today is entirely different from what it was in the years preceding the dissolution of the Standard Oil Company. The contrast between the concentration shown here in Exhibit One and the

---

[8] Simon N. Whitney, *Antitrust Policies, American Experience in Twenty Industries,* (New York: The Twentieth Century Fund, 1958), p. 101.

[9]Thomas G. Manning, E. David Cronon and Howard L. Lamar, *The Standard Oil Company: The Rise of a National Monopoly,* (New York: Henry Holt and Company, 1960), p. 43.

[10]Harold F. Williamson, Ralph L. Andreano, Arnold R. Daum, and Gilbert C. Klose, *The American Petroleum Industry: The Age of Energy 1899-1959,* (Evanston: Northwestern University Press, 1963), p. 9.

current low concentration illustrated by Exhibit 3 on page 34 demonstrates clearly that Standard Oil's imposing dominance of the petroleum industry prior to 1911 has no present-day parallel.

Production

Standard's control over crude supplies ranged from 92% in 1880 to 70% to 80% of older fields and 10% to 30% of newer fields in the West in 1911. Standard Oil did not actually dominate production of crude oil; rather, because of its purchasing power, and its dominant position in pipelines, it was able to set the prices of the crude oil it purchased.[11]

Today, (1974) the largest producer of crude oil produces just 9% of the total. And neither the top 4 companies (26%) nor the top 8 companies (42%) can match Standard's pre-1911 position.

Refining

Standard's share of refining capacity ranged from 90% to 95% in 1880 to 64% in 1911.

By comparison, the top refiner today accounts for only 8% of capacity. The top 4, 31%; the top 8, 54%.

Transportation

Standard Oil had almost a total monopoly over pipeline transportation. Through its relationships first with railroads and then with pipelines, Standard was able to transport crude oil at a lower cost than its competitors.[12]

The top company in pipeline ownership by volume has only 10% of interstate pipelines. The top 4, 34%; the top 8, 55%. And tankers are even less concentrated.

[11]Federal Trade Commission, Report, *Petroleum Industry Prices, Profits, and Competition*, Senate Document 61, 70th Congress, 1st Session, 1928, p. 64.

[12]Bentley, *Op. Cit.*, pp. 10-33.

Marketing

At the peak of its domination, Standard sold 90% to 95% of kerosene sold in 1880. By 1911, Standard still sold 75% of the kerosene, and 66% of a relatively minor product called gasoline.

In the gasoline market today, the top company accounts for 8%. The top 4 sell 30%; the top 8, 52%.

Ownership

In 1900 , John D. Rockefeller owned 42.9% of Standard Oil of New Jersey. Fifteen other individual stockholders accounted for an additional 39.5%. This meant that over 80% of the company which virtually dominated the petroleum industry was owned by only 16 individuals.[13] In 1911, ten men still owned 37.7% of Standard's stock, with 24.9% held by Rockefeller. At the time of its dissolution, all of the stock of the Standard Oil Company was held by only 6,000 stockholders.[14]

In contrast, today the shares of just the six largest oil companies are owned by 2-1/2 million direct shareowners and another 11-1/2 million indirect owners. In other words, 14 million Americans, or about 6.5% of the population, are shareowners of just the six largest companies compared to 6,000, or only about 1/10,000 of one percent of the population in 1911.

[13] *Ibid.*, p. 65.
[14] "Standard Is Strong in Disunion," Business Week, June 22, 1946.

Profits

Economic historians have indicated that Standard Oil's profit rates were twice that of profit rates in general during the years leading to up 1911. Standard's profit as a percentage of net worth was 23.1% in 1906; 20.7% in 1904; 27.0% in 1902; 27% in 1900.[15]

During 1975, the 25 leading oil companies had a comparable profit rate of 13.5, about half the Standard rates the decade prior to the breakup. For the ten years 1965-1974, the average profit as a percentage of net worth on petroleum companies was 13.4%. The ten year average for all mining was 14.7%; for all manufacturing, 13.0%.[16]

Summary of Standard Oil's Position in the American Petroleum Industry: 1880-1911

Percent of control over crude oil supplies

Fields	1880	1899	1906	1911
Appalachian	92	88	72	78
Lima-Indiana		85	95	90
Gulf Coast			10	10
Mid-Continent			45	44
Illinois			100	83
California			29	29

Percent of control over refinery capacity

	1880	1899	1906	1911
Share of rated daily crude capacity	90-95	82	70	64

Percent of major products sold

	1880	1899	1906-1911
Kerosene	90-95	85	75
Lubes		40	55
Waxes		50	67
Fuel Oil		85	31
Gasoline		85	66

Source: Williamson and Andreano, "Competitive Structure of the Petroleum Industry," *Oil's First Century*, 73.

---

[15]Manning, et al., *Op. Cit.*, pp. 32-33.

[16]First National City Bank, *Monthly Letter*, April 1975.

The Breakup of The Standard Oil Company (N.J.)

On May 15, 1911 the Supreme Court handed down its decision that the Standard Oil Company of New Jersey consituted a monopoly in restraint of trade. The Court decreed that the Sherman Act outlawed unreasonable monopoly and that the Standard organization was an unreasonable monopoly. The Justices concluded further that the Standard Oil Company intended to establish a monopoly and "to drive others from the field and exclude them from their right to trade."

Standard Oil was given six months to accomplish the divorcement of 33 companies. The stocks of the various soon-to-be-separate companies had to be distributed on a pro rata basis to the 6,000 shareholders of Standard Oil. Standard Oil lost 57 percent of its net value and 91 percent of its annual earning power.[17]

The Effect of the Breakup of the Standard Companies

The effect of the dissolution on the Standard organization was immediate. John D. Rockefeller resigned as nominal president of Standard of New Jersey. William Rockefeller and his son, William G. Rockefeller, also stepped down from their positions in the company.[18]

The dissolution into separate organizations was accomplished with little difficulty, as the various companies were already operating with their own management.[19] It was also thought that Standard's management had raised the capitalization of its subsidiaries to facilitate divestiture.[20]

While the technical task of creating a number of companies out of one was easily accomplished, the goal of creating viable companies capable of competing effectively in the oil industry required years to effect.

Thus, during the era following the 1911 case, many of the Standard Companies integrated forward to handle surpluses and integrated backward to handle shortages. Vertical integration was the method by which many "Standard" and "independent" companies achieved economic success during the period between the World Wars.[21]

While there was a period following the dissolution in which the old Standard Companies, almost as if by gentlemen's agreement, appeared to

---

[17]George Sweet Gibb and Evelyn H. Knowlton, *history of the Standard Oil Company (New Jersey): The Resurgent Years, 1911-1927,* (New York: Harper & Brothers, 1956), p. 7.

[18]Nevins, *Op. Cit.*, Vol. II, p. 380.

[19]*Ibid.*, p. 379.

[20]*The New York Times,* May 16, 1911, p. 4.

[21]John G. McLean and Robert Wm. Haigh, *The Growth of Integrated Oil Companies,* (Boston: Graduate School of Business Administration, Harvard University, 1954), p. 668.

be less than rigorous in their competition with each other, this period did not last.[22]

Today there is no precise dividing line between integrated and non-integrated companies. It is more realistic to view oil companies on a continuum with those on one end almost totally integrated and those on the other performing only a single function.[23]

How Did the Shareowners Fare?

For some, the motive behind dissolution of Standard Oil had been a desire to see a truly competitive situation in the petroleum industry. Many businessmen, particularly independent oil men, were among those who saw the destruction of the Standard Trust as a positive development for the nation. For some, however, the main concern was that the concentration of wealth and power in Standard Oil was bad per se. While the company was indeed broken up, two factors were discouraging to Standard's critics who were primarily antibusiness.

The first was that while the companies were now compelled to compete with one another, at first it appeared that this "competition" was on paper only. After all, the same people owned the same properties. One well publicized result of the dissolution was that one prominent Standard vice-president merely changed titles, and moved to a new office a few feet down the hall at 26 Broadway.

The second disappointment to many critics of business was that while the management of Standard was broken up, and the dominant position of Standard was clearly gone, *the owners*, John D. Rockefeller among them, appeared to get even richer as a result of the decree. Dividends for holders of Standard's $100-par stock were reduced from $37 to $20[24] but the price of Standard stock and of the other Standard Companies soon began to rise, as dividends increased. Standard of New Jersey had paid its highest dividends, 48 percent, in 1900 and 1901. The rates in the years before dissolution had ranged between 36 percent and 45 percent. During 1912, the first year following dissolution, 26 of 34 Standard Companies paid dividends amounting to 53 percent of the outstanding capital stock of the old Standard of New Jersey.[25]

During the 4-1/2 months of 1911 prior to the Supreme Court decision of May 15, Standard's stock had risen 61-1/4 points to 679-3/4 on the day of

---

[22]Nevins, *Op. Cit.*, Vol. II, p. 381. *Business Week, Op. Cit.*,

[23]McLean & Haigh, *Op. Cit.*, p. 17.

[24]"Standard Oil Co. (N.J.)," *Fortune*, n.d., (1938), p. 149.

[25]Nevins, *Op. Cit.*, p. 382.

the decision itself. And the stock was 94-3/4 higher than at its lowest point in 1910.[26]

The following table shows the increases in selected oil company stocks between January and October of 1912:[27]

Company	January	May 15	October
Standard (N.J.)	360-375	384-385	590-595
Standard of N.Y.	260-275	395-405	560-580
Atlantic Refining	260-300	375-395	610-620
Galena-Signal	215-224	225-235	240-245
South Penn Oil	350-375	600-620	800-825
*Texas Company	80-88	98-99	120 1/2-126 3/4
*Houston Oil	8-9 1/4	10 3/4-11 1/4	17-23

*Independent

While the rise in the Standard stocks has been cited as a beneficial result of the Standard Companies, breakup, it should be noted that the Texas Company and the Houston Oil Company, both independent companies that were competing with Standard prior to the breakup, were also enjoying substantial increases in the price of their stocks. Thus, economic conditions appear to have been generally favorable for all oil companies.

Why the Oil Boom After Dissolution?

Three factors are generally cited as the reasons for the explosive progress in the petroleum industry during the 20 years immediately following dissolution of the Standard Trust.

The first fact, and clearly an important one, was the growth of the automobile industry in the United States, and the accompanying growth in the gasoline market. Henry Ford's decision to manufacture automobiles for the masses had tremendous implications for the oil industry. In 1909, 126,593 motor vehicles were manufactured in the United States. In 1914, 569,054. In 1919, 1,683,916.[28] Prior to 1911, the major petroleum product was kerosine for use as an illuminant. After 1911, the major product was gasoline in a rapidly expanding market.[29] The number of automobiles registered in 1900 was 8,000; 1910, 458,000; 1915, 2,491,000; 1920,

---

[26] *Times, Op. Cit.*, p. 4.

[27] Nevins, *Op. Cit.*, p. 382 *Wall Street Journal*, Jan. 1, 15; May 15; Oct. 2 and Nov. 11, 1912.

[28] Williamson,, et al., *Op. Cit.*, p. 190.

[29] McLean and Haigh, *Op. Cit.*, p. 57.

9,239,000; 1925, 19,941,000.[30] In addition to the sheer volume of sales generated, an entirely new marketing system had to be developed geared toward the automotive public. Forward integration was a widespread result.

A second factor behind the petroleum industry growth after 1911 was the coming of World War I. Lloyd George commented that: "The Allies floated to victory on a sea of oil."[31] Whereas serving the gasoline market created strong trends toward forward integration, crude shortages occasioned by the War caused many refining companies to integrate backward into production.[32]

Standard of New Jersey, for example, which supplied almost half of France's petroleum imports during the war, saw its exports of gasoline increase 14 percent in 1917, another 33 percent in 1918. Fuel oil exports increased 221 percent in 1917, another 81 percent in 1918. Standard's sales to the U.S. Navy from Atlantic ports went from 4,656 barrels in July of 1917 to 368,883 barrels in May of 1918.[33]

While all of the companies suffered severe losses in men, materiel, and money during the war as the Germans sank tankers supplying the Allies in Europe, clearly the War brought on an expansion to the industry.

The third factor in the growth of all oil companies following 1911 was the discovery of large new crude oil fields. Discovery and especially rapid production of new fields had a significant effect on the market, especially before prorationing. While there may be hundreds of producing fields at any one time, up until 1951, for example, only 25 fields had produced 29 percent of the crude output up until that time.[34] Crude production soared from 220.5 million barrels in 1911 to 378.4 million barrels in 1919.[35] Demand—what with the gasoline market and World War I discussed above—kept pace. The cost per barrel was $ .61 in 1911; $2.21 in 1919.[36] The new discoveries in the West not only helped the industry grow, but also encouraged diversity in its growth as new companies sprang up, and as old independents began to meet the old Standard Companies in size, etc.

---

[30]*Ibid.*, p. 267.

[31]Nevins, *Op. Cit.*, p. 384.

[32]McLean and Haigh, *Op. Cit.*, p. 668.

[33]Gibb and Knowlton, *Op. Cit.*, pp. 224-225.

[34]McLean and Haigh, *Op. Cit.*, p. 82.

[35]Williamson, et al., *Op. Cit.*, p. 37.

[36]*Ibid.*, p. 39.

Conclusion

The breakup of the Standard Oil Company of New Jersey by the Supreme Court in 1911 was a response to a particular company in a particular setting and the Court was applying a criminal antitrust statute that had general applicability. The economic conditions that led up to the 1911 decision were entirely different from those prevailing today. Whereas Standard Oil dominated virtually the entire oil industry from the decade of the 1870's to 1911, today the concentration ratios of neither the top 4 nor the top 8 companies can match Standard's earlier dominance.

In addition, wide shareownership of today's oil industry (14 million direct and indirect shareowners of the top 6 companies) contrast sharply with 16 individuals owning 80 percent of Standard Oil, or only 6,000 shareholders in total.

Another contrast is present in the high profit rates of Standard during the decade prior to dissolution compared to oil company profit rates today generally in line with other industries.

It is also apparent that the *results* of the dissolution would be quite different under present conditions. Legislation to require the vertical disintegration of oil companies would prevent use of the most effective method that oil companies used to achieve economic viability during the decades following dissolution.

6

Vertical Divestiture: Exploration and Production

TESTIMONY BY L. C. SOILEAU III, PRESIDENT, THE CALIFORNIA COMPANY DIVISION OF CHEVRON OIL COMPANY, BEFORE THE SUBCOMMITTEE ON ANTITRUST AND MONOPOLY, COMMITTEE ON THE JUDICIARY, UNITED STATES SENATE — NOVEMBER 19,1975

Summary: Divestiture would reduce exploration and competition as companies approaching limits set by the bill would hold back on activities to make sure they do not fall under the provisions of the bill. Financing would be more costly. Efficiencies and economies of centralized management and services would be lost. Production has not been held back. Economic, physical and legal considerations make shutting-in production unattractive. Further, the integrated companies need the crude to supply their refining and marketing operations. SOCAL's income from foreign operations has been used to finance its domestic energy program. If these were split off, less income would be available for U.S. exploration and production investment. The charge that independent producers are at the mercy of majors owning pipelines is untrue. The most persuasive evidence against it is the absence of complaints by independent producers.

L. C. Soileau III speaks . . .

I am L. C. Soileau III, President of The California Company Division of Chevron Oil Company. My oil field experience dates back to 1945 when I went to work for The California Company as a roustabout. I have worked as a tool pusher, a construction engineer and a drilling engineer. I participated in the drilling of our first offshore well in the Gulf of Mexico in 1948. I have had experience in drilling and producing in Venezuela, the Rocky Mountains, California, Alaska and the Gulf Coast area offshore and onshore.

The purpose of my appearance before the Subcommittee is to discuss how S.2387 would affect the exploration and production segments of the oil and gas business.

The purpose of our business is very simple—to find and produce as large a supply of petroleum as possible, and at as low a price as possible. I believe the history of the petroleum industry in America shows that the integrated oil companies have an outstanding record in both respects. I believe the facts will also show that divestiture not only would not produce better results, but would bring about a reduction in supplies and higher prices.

I will confine my comments essentially to the prospective effects of S.2387, or similar legislation, on the availability ad price of domestic petroleum. Throughout my statement the terms "integrated" and "non-integrated" will be used. These terms refer to a method of operation, and are not to be confused with "major" and "independent," which refer to size. An "integrated" company may be either a "major" or an "independent." I might stress that some of the successful smaller oil companies in business today are integrated companies.

The Subcommittee has been told by members of its staff that "there is no evidence that vertical integration has made any significant contribution to efficiency," and that "vertical divorcement is not a drastic measure."

These statements reflect a lack of understanding of how an integrated company conducts its business, and a total indifference to the clear and inevitable consequences of divorcement.

So that there be no misunderstanding concerning the nature of exploration and production, and how vertical integration facilitates these processes, I have described in some detail in Exhibit 1 how we go about finding and producing oil and natural gas.

WHY DIVESTITURE WOULD ADVERSELY AFFECT THE SUPPLY AND RAISE THE PRICE OF PETROLEUM

Divestiture Would Not Lower the Price of Oil

In all this discussion one point must be kept in mind: It is now clear beyond dispute that our country will never again enjoy the luxury of "cheap oil." Production from oil deposits discovered in readily accessible areas "peaked" in the late 1960's, and is now rapidly declining.

New domestic supplies will be higher in cost because our future supplies will come primarily from five sources: (1) foreign oil; (2) OCS production from expensive wells in deep water; (3) onshore production from deep reservoirs that are costly to drill and operate; (4) Alaska; and (5) assisted recovery methods applied to fields in long-established producing areas. All of these are highcost sources.

Certainly a significant price decrease in the cost of oil cannot come out of the net income of the larger integrated companies. My own company, for example, had total U.S. earnings from all operations (excluding chemicals) of 1.4 cents per gallon of U.S. refined oil sales over the ten-year period of 1965 through 1974.

Even in the unusual year 1974 this 1.4 cents went up only to 1.8 cents per gallon. In the first nine months of 1975 our earnings per gallon dropped below the prior ten-year average to 1.3 cents per gallon. Other large integrated firms have a cents-per-gallon net income of similar magnitude.

The avowed purpose of S.2387 is to reduce the price of oil by divestiture of the larger integrated companies. Confiscating the entire net income of these companies thus could not have a large effect. This underlines the fact that a reduction in oil company earnings—even a 100 per cent reduction—could have only a negligible effect on the price of petroleum products.

The relationship of profits to investment is almost always avoided by people attacking the oil industry. They point to the industry's profits in terms of a large number of dollars, but they seldom acknowledge the tremendous investment it takes to generate those profits, nor do they compare our rate of return with that of other industries. And it is the rate of return—the number of cents of profit for each dollar invested—that is the key. The rate of return in the petroleum industry is already less than in American business generally. If, in some way that is not clear now, profits were forced even lower, how could capital be available for the immense amount of expansion which is needed to meet the nation's energy demands?

No Monopoly Exists.

It has been charged that the integrated majors monopolize crude production and that this so-called monopolistic domination of production is somehow related to or extended by vertical integration into transportation, refining and marketing.

Let's look at the evidence: First, the figures should prove to anyone that there certainly is no monopolistic control at the production level. Total U.S. production of crude oil and natural gas liquids in 1974 was over ten million barrels per day. This oil was produced by literally thousands of separate companies and private individuals.[1]

[1]See Exhibit 3, page 34.

Looking at natural gas production, much the same situation exists, except that the percentages accounted for by the four, eight and twenty largest companies are even smaller.

As for the Outer Continental Shelf, 61 per cent of crude oil and natural gas liquids production and 53 per cent of natural gas production in 1974 were attributed to the eight largest integrated majors by the FTC's Bureaus of Competition and Economics in their October 1975 report to the Commission.

This concentration, which is higher than the overall figures because the money and expertise needed to go offshore have generally been found in the larger companies, has been rapidly decreasing. The crude and natural gas liquids production of companies other than the eight largest has increased dramatically from 9.6 per cent in 1965 to 39.4 per cent in 1974—an increase of 310 per cent over the past ten years. Similarly, in natural gas production, their shares have increased over 400%, from 11.7 per cent in 1966 to 47.1 per cent in 1974.

Between 1960 and 1970 the number of companies actively participating in OCS sales more than doubled. In 1962, the smaller semi-integrated and independent petroleum companies won just over one third of the lease acreage offered. In 1972, the share of acreage awarded to companies other than the eight largest was about two thirds.

Monopolistic control cannot exist where entry into the marketplace is not difficult. Entry of new companies into exploration and production is not difficult. Privately-owned oil and gas leases are available to anyone. Both Federal and state governments regularly hold bidding auctions where oil and gas leases are sold to the highest bidder. All that is required is the funds to enable the new entrant to finance lease purchases and subsequent exploration and drilling costs. In expensive OCS areas the joint bidding mechanism is available to small independent producers.

As I understand it, in law and economics a major determinant of the existence of monopoly power is the ability to restrict production and/or raise prices to levels where abnormally high profits can be realized. No accepted economic theory in my view could support the contention that at the low levels of concentration shown above, monopoly power could exist.

There Are No Monopoly Profits in OCS Production

OCS production has not been and will not be a source of alleged "artificially high" crude profits for the industry.

As for the past, a study prepared for the Department of Interior in 1974 by Professors Edward W. Erickson and Robert W. Spann concluded that the average winning bids in the 1972 and 1973 lease sales "were consistent with an eight per cent or less return on total capital expenditures given (then) reasonable expectations concerning future petroleum prices, drilling costs and the probability of commercially successful production."

Such a rate of return is not even adequate, let alone excessive. As a matter of additional interest, this same study also concluded that the much-criticized practice of joint bidding not only was not anti-competitive, but that it acted as a risk-sharing device, and probably facilitated the entry of small and medium-sized firms.

I invite your attention to the following conclusions from an October 1975 report to the Federal Trade Commission by its own Bureaus of Competition and Economics:

> "The market for offshore oil and gas leases appears to be effectively competitive. We have found no evidence of collusion (which would be difficult in view of the large number of bidders).
>
> * * * * *
>
> "Evidence on offshore petroleum operations indicates that they have on average not yielded above-normal profits. To the extent this is true, it supports an inference that bonus bidding offshore has been effective in capturing the economic rent from the tracts sold."

A study was made recently by a major integrated oil company of OCS leases in the Gulf of Mexico upon which discoveries were made during the period 1964-1973. The purpose of the study was to analyze the distribution of past and future revenues from these leases. Prices that actually prevailed from 1964 to October 1975 were used for production.

It was estimated in the study that four billion barrels of oil and 24 trillion cubic feet of gas have been discovered on the leases between 1964 and 1973, which will produce a total revenue of $35 billion. After deducting $11 billion in exploration, development and operating costs, some $24 billion in net revenues will be left. Eighteen billion dollars of these net revenues will be paid to the United States—$9 billion for bonuses and rentals and $9 billion for royalties and taxes. Six billion dollars will remain for the lease operators.

Out of the net revenues the Federal government will therefore receive three times as much as the oil industry. The industry rate of return, after taxes, based upon discounted cash flow analysis, is projected to be only about five per cent.

Incidentally, one important reason for the low rate of return on offshore operations is the relatively long period of time that normally elapses between the acquisition of a lease and the date of first production from the lease. In the experience of my company this period of time is typically five to eight years.

Divestiture Would Reduce Exploration and Production Activity

One who has had experience in exploration and production must reach the strong conviction that enactment of S.2387 would be followed by a significant decline in oil and gas exploration in the U.S.

This slump would occur initially in the three-year period following enactment of the statute. During this time the industry would be required to prepare and to effectuate plans for divestiture. Literally millions of daily transactions essential to the stable and steady flow of oil from wells to gasoline pumps to industrial plants and utilities, and homes—indeed, the whole fabric of relations needed to keep our huge economy going—would be torn up. Obviously this would be a very difficult period, and one of great uncertainty, and I believe companies would trim their exploration and producing programs until they saw how things would work out.

For an indeterminate interval of time after the end of the three-year period, I believe there would also be a natural reluctance to take major risks.

In addition to disrupting the activities of those companies immediately affected by this bill, three features of the bill would directly reduce future competition.

1. The arbitrary limits placed on the size of activities which are allowed without being subject to divestiture would be reached in the years to come. Companies in this position could be expected to hold back on their exploration and production or refining or marketing expansion activities, as the case may be, to make sure they did not come under this provision. In effect, they would cease to be fully competitive as they approached the limits the bill sets forth.

2. This bill would prohibit any "petroleum transporter" from entering into competition as an explorer or producer in the future. Once this bill went into effect, all companies which qualified as "petroleum transporters" would be unable to join in the competition for gas and oil supplies.

3. New companies would not be likely to begin business with large activities (say, a 210,000 barrel per day refinery) because they would be prevented by this bill from entering into competition in any of the other parts of the business.

Thus this bill, which presumably is intended to encourage competition, would have exactly the opposite effect.

OTHER MATTERS OF IMPORTANCE AFFECTING EXPLORATION AND PRODUCTION

Financing Would Be More Costly

Expenditures of large sums of money obviously must be made to conduct geophysical surveys, to buy leases, to drill wells and to construct drilling and production facilities. When borrowing is necessary, the integrated company, with its large and diversified assets and greater stability of earning power, can usually do so at lower interest rates than a non-integrated exploration and production company, all of whose activities are inherently high-risk.

By denying to exploration and production the financial advantages of the integrated company, S.2387 would result in exploration and producing companies having to pay higher rates of interest for loans, thus increasing the cost of producing crude oil and natural gas.

Research and Development Would Be Retarded

A vertically integrated organization can justify the financial support of a substantial, continuing research effort, the economic benefits of which can be extended over the various functions which that organization performs.

Only through the efforts of an efficient vertically integrated oil industry has it been possible in the past to meet the energy demands of the nation with the types of crude oil available for processing. These include the now abundant high sulphur Mid-East crudes, high sulphur asphaltic Venezuelan crudes, and waxy domestic and foreign crudes.

The great technological advancements of the petroleum industry, with their corresponding benefits to the public, have been primarily the result of a coordinated research and development effort carried out by integrated firms. For example, it was well over ten years ago that my company embarked on a strong and expensive program of research on refinery processes for desulphurizing and converting to light products the high sulphur, light and heavy Mid-East crudes that we predicted would be a principal source of supply. This effort has proved invaluable in meeting today's demand for low sulphur content products, as well as other air pollution requirements.

Research and development relating to new exploration tools, secondary and tertiary recovery, and improved production methods, have helped to supplement the diminishing supply of domestic crude oils, thus reducing our shortfall and consequently our dependency on imports.

Close coordination and knowledge of producing, refining and marketing sectors of the industry is essential to anticipate future needs and to direct research and development toward the proper goals in each of these sectors. This can best be achieved in the vertically integrated structure of the major oil companies.

Risk Taking Would Be Inhibited

Because the integrated company generates income from sources other than the production of crude oil and natural gas, it is usually able to withstand the shock of a run of bad luck in its exploratory program. The continuity of a promising program is not endangered by the drilling of early dry holes and wide diversification of exploratory effort is feasible.

On the other hand, the non-integrated exploration and production company contemplated by S.2387, whose funds are derived from limited sources, obviously would not have the same financial flexibility and could not tolerate a low rate of return for any extended period of time. The choice that the non-integrated company would have to face is to abandon prematurely its exploratory and development efforts, or to curtail these efforts, or to run the risk of insolvency. Thus the search for new reserves of oil and gas would necessarily be seriously impaired if S.2387 were adopted.

Management and Manpower Would Be Used Less Efficiently

Management decisions of an integrated company are made with access to information about the company's operations in all operating functions. Because this data is available, the company's various activities can be synchronized and coordinated, and an efficient operation throughout the entire process is facilitated.

Centralization of specialized manpower in the integrated company below the management level also adds to efficiency. Accountants, landmen, lawyers, supply specialists, purchasing agents, personnel specialists and others perform services for all of the major operating divisions of the company.

Each of the non-integrated companies contemplated by S.2387 would presumably have its own management and its specialized manpower. Thus there would be a multiplication of personnel. The efficiencies and economics of centralized management and services would be lost.

By my prior comments I do not mean to imply that vertical integration is best for all oil companies. Individual company situations will differ. There is room and need for both integrated and non-integrated companies in our industry. The growth and continued success of both demonstrate that there is no single "right" way to operate in the oil business.

But there is a right way to choose the best mix. The marketplace, if left alone, will determine the optimum mixture of integrated and non-integrated companies.

Integrated Companies Have Not Held Back Production

In recent weeks we have heard the claim that crude oil and natural gas production in the United States is being deliberately held back because the oil companies anticipate higher prices when price controls are removed.

This charge simply is not true.

Socal, for example, is producing every barrel of oil and gas it can economically, safely and legally produce in the United States. On October 31, 1975 Mr. J. A. Babin of Shell Oil Company gave an excellent series of illustrations about what is meant in this context by the word "economically." I therefore will not dwell extensively on the point today, but will summarize briefly what he said, and what I mean.

When I say we produce every barrel we can economically produce, I mean we expend resources to produce oil so long as the incremental profit from these activities can be expected to provide an adequate return. In deciding whether to continue producing from existing facilities, we must be satisfied that the value of the recoverable oil to be produced will exceed the operating costs required to produce it. In deciding whether to invest in capital items such as wells and platforms, an adequate return must include a return on investment sufficiently attractive to justify the expenditure in question in comparison to other possible uses of those resources. Otherwise, the resources will not have been used wisely.

The very nature of oil and gas reservoirs limits production rates, and rules out claimed incentives to hold back oil. Most significantly, oil and gas production rates generally decline as reserves are depleted—at an average of about ten per cent per year in the U.S.

A field, if shut-in and subsequently returned to production, will normally quickly stabilize at its previous rate. It will then decline from that level in the same pattern as if production had not been interrupted. The production rate cannot be jumped after a shut-in to catch up on the deferred production. Production that is deferred can be recovered only gradually over the remaining life of the field.

Exhibit 3 gives an example showing that it takes seven years to recover one half of the production lost if a field is shut-in for one year.

Oil and gas in a reservoir are not like wheat in storage. Wheat can be held for a year and sold, together with next year's crop at next year's price.

Shutting-in crude oil production causes an immediate loss of income which can only be recouped over a long period.

There are also some other significant factors which deter shutting-in production:

1. Unless wells on nearby leases are also shut-in, oil and gas may be drained from the shut-in wells. If nearby wells are owned by other parties, the operator of the shut-in wells and his landowner-lessor will probably never be able to recover the volume that is drained.

2. Practically all oil and gas leases require the operator-lessee to produce and to market oil and gas with reasonable diligence. Arbitrarily shutting-in production would be a violation of this lease obligation and could result in cancellation of the lease.

3. Some wells, if shut-in for a significant time, do not return to their previous production level, when reopened, for a variety of reasons. For example, they may become plugged with sand.

Economic, physical, and legal considerations thus make shutting-in or cutting-back production unattractive in the vast majority of cases.

It should be noted that if there were any substance to the charge that oil companies have an economic incentive to withhold production, this incentive would apply more strongly to the *non-integrated* producers, since they would not need the crude immediately to supply their refining and marketing operations.

Following the completion of my company's current expansion of refining capacity, our crude oil needs will be more than three times our current domestic crude production. In other words, approximately two thirds of the crude required to utilize all our refining capacity will have to be purchased in the U.S. or imported at the world market price.

Dr. Measday has suggested to this Subcommittee that Federal leases in the Gulf of Mexico could produce almost twice as much as they are now producing. He bases this suggestion on the fact that these leases have been producing below the maximum efficient rates (MER) approved by the U.S. geological survey under OCS Order No. 11. He concludes that the lessees "have determined that it is not in their economic interests at this time to develop the leases."

In his October 31 testimony the subcommittee heard Mr. Babin of Shell refute Dr. Measday's charges. I have read the statement submitted by Mr. Babin, and I consider it to be a realistic explanation of why actual production is below MER. In determining MERs under OCS Order No. 11,

Chevron uses basically the same methods that are described by Mr. Babin.

Like Mr. Babin, I believe that actual production in the Gulf of Mexico is less than MER for several reasons:

1. MERs for some reservoirs are fixed initially at levels that are high enough to accommodate new production from wells that are subsequently drilled or reworked. Thus, there is an initial built-in gap between MER and actual production.

2. Wells normally decline in productivity. Even though a reservoir could produce its MER at the moment the MER is determined, the reservoir loses its ability to produce that MER within a very short time. Thus, a MER fixed for a reservoir under OCS Order No. 11 in September 1975 probably cannot be produced today because of the reservoir's normal decline in productivity.

3. Production is normally interrupted when additional wells are drilled, when existing wells are reworked, when production facilities are repaired, when work is being done to comply with new Federal regulations such as safety regulations, when precautions are taken against bad weather, and for other reasons. During these shutdowns the gap between MER and actual production widens.

I believe that actual production in the Gulf of Mexico is less than MER because of the cumulative impact of the factors set forth above. I can assure you without hesitation that this is true in the case of Chevron-Calco.

There Has Been No Relaxation in the Efforts of Integrated Companies to Find New Domestic Reserves

Some persons have claimed that the integrated oil companies have not tried hard enough to find domestic oil. Apparently they base this charge on the drilling statistics provided by the American Association of Petroleum Geologists which show that, of the integrated companies, the 16 largest oil producers are "operators" for only about ten per cent of the total number of new field wildcat wells drilled.

Actually AAPG's most recent information which takes into account the size of discoveries, clearly demonstrates that all segments of the exploration industry contribute their vital shares to oil and gas discoveries. As shown in Exhibit 4, the 16 largest companies were "operators" of 25 per cent of the "significant" new field wildcat discovery wells. However, their discovery well success ratio was several times that of the rest of the industry.

Further, their ratio of new reserves found per wildcat well drilled is many times that of the rest of the industry—an undeniable indication of their effectiveness in bringing advanced technology to bear in finding major new supplies of oil and gas. As a result, their contribution to estimated new actual reserves was more than 50 per cent of the total for the past six years. This figure would be higher if it included their interests in wells which were drilled by others.

Confronted with these facts, no one can say that the large integrated companies are not doing their part in finding oil in the United States.

In support of this view, I invite you to examine the record of Standard Oil Company of California. During the ten-year period 1965-1974, Socal invested $1.7 billion in *searching for* domestic oil and gas, and $1.6 billion in wells, platforms and other capital items for the purpose of *producing* domestic oil and gas. This $3.3 billion total exceeded by $1.2 billion the total U.S. profits of the company from *all* of its integrated operations during that period. It is almost twice the amount we spent in foreign countries in the search for and development of oil and gas in the same period.

In the past six years, 1970 through 1975 to date, Chevron-Calco has participated in all 14 Federal lease sales in the Gulf of Mexico, and bidding alone or with partners, has submitted 460 bids. These bids totaled $3.6 billion, of which Chevron-Calco's share was $1.74 billion.

Chevron-Calco alone, and Chevron-Calco and its partners, were awarded 146 leases during this period, for which bonuses were paid totaling $1.279 billion. Chevron-Calco's share of these bonuses was $636.5 million.

Of further significance, in eight of the last ten years most of Standard Oil Company of California's income has come from operations abroad. In 1975 profits from the company's foreign operations have been used to finance U.S. energy programs, because domestic earnings continued to be depressed by restrictive price controls and loss of the oil depletion allowance. If these foreign operations were split off into another company, the income stream available for domestic exploration and production investment would be seriously reduced.

Independent Producers Have Ample Access to Gathering Lines

During the hearings of this Subcommittee, and elsewhere, comments have been made to the effect that the small independent producers are handicapped in not being able to get their oil from the wellhead to the point of delivery to the ultimate purchaser, without utilizing a gathering line owned by a major or a couple of majors.

References have been made to the fact that title to the crude is transferred to the owner of the gathering line before shipment and returned to the producer at the destination. It is said that this practice puts the small producer at the mercy of the pipeline owner and is thus anticompetitive.

This charge represents a complete distortion of a simple transportation arrangement which has been adopted to facilitate the movement of crude oil for small and large producers alike.

The shipper and pipeline operator simultaneously agree on sales and repurchase terms. The shipper knows before he puts oil in that he will get oil out, just as in a tariff arrangement. The practice of crude purchases at one end of the line and resales at the other occurs because different shippers have different quality crude oils and the pipeline owner ordinarily has no facilities for segregation during shipment.

For example, the shipper frequently puts into the line crude of a different gravity and value than he receives from the "common stream" at the other end. The purchase and resale is a simple method of adjusting for these differences in value or location.

This arrangement is the functional equivalent of continuous ownership by the independent producer and payment of a transportation fee or tariff. The advantages to the shipper are several, including:

1. Neither the shipper nor the pipeline owner is required to build unnecessary tank storage at the point of receipt or delivery. Thus, capital investment and other costs are avoided—to the benefit of all parties, including the ultimate consumer, for the money that is not spent can be used to find and produce more oil.

2. The independent producer enjoys greater flexibility with respect to the date the crude is to be returned.

Dr. Measday alleged that the Placid group of producers was forced to sell their offshore production to Chevron with no option to receive their oil production in kind. The facts of the matter are completely opposite to those presented by Dr. Measday. On November 1, 1969, Chevron entered into an agreement to purchase South Timbalier Block 179 oil production from a group of eight producers (including Placid) and sell this production back to General Crude, operator for the group. Point of purchase was Chevron's pipeline in South Timbalier 189, and point of sale was Cal-Ky's Empire Terminal.

On December 1, 1970, Placid replaced General Crude as operator for the group. By letter dated January 21, 1972, Placid requested that we waive Placid's obligation to repurchase the group's oil production. Effective May 1, 1972, the repurchase obligation was suspended until January 1, 1973, and continued in suspension thereafter on 30 days' notice, but notice was never given by either party. Incidentally, this property ceased production by May of 1974. Phil Sowards, Manager of Crude Oil Sales for Placid, verbally confirmed that Placid had no need for the oil production and was quite willing to sell to Chevron and the procedure for handling the crude oil was to their liking and advantage.

Perhaps the most persuasive argument against the charge that the present system is unfair and inequitable is the absence of complaints by independent producers to the owners of the gathering lines or to appropriate Federal authorities. Some witnesses have attempted to explain this absence of complaints on the ground that small shippers are reluctant to antagonize pipeline owners. This explanation has no factual support. In today's climate there simply is no one who hesitates, out of any fear of economic reprisal, to take advantage of an opportunity to complain about the practices of anyone else in our society.

The real reason for the absence of complaints is that the charge is untrue. Socal regularly carries oil for other shippers on its gathering and mainline systems, for the most part under buy-sell arrangements. Other companies likewise use our terminal facilities, such as the Expire Terminal in Louisiana. I believe this to be true of the industry in general. Those who contend otherwise simply do not know the facts.

Joint Ventures Do Not Reduce Competition

Let us look now at the charge that the oil industry is unique in the number of its "cooperative" ventures, such as joint bids on OCS leases, jointly owned pipelines, joint ventures abroad and crude oil exchanges.

Critics argue that two adverse consequences follow from these relationships:

1. Each company learns about the internal affairs of the other participants, which information weakens the vigor of competition; and

2. Competitors who cooperate in one area cannot be expected to compete in other areas.

As to the first part of the argument:

What one company learns about another as the result of participation in joint producing facilities, or a jointly-owned pipeline, is information that is either publicly available or of no competitive significance. Joint bidders at OCS sales do not discuss or disclose competitive aspects of their activities or plans in respect of areas outside of the joint venture. Socal has bid jointly with over 50 separate companies in Gulf of Mexico sales. Most of these bids were not with other integrated companies. More importantly, we submitted bids on over 300 tracts which were competitive with our partners' bids outside of the joint bid areas. I know of nothing as competitive as bidding 100 million dollars on a single tract in a sealed bid OCS sale.

As to the second part of the argument:

Senator Tunney has questioned whether Socal and Texaco could be expected to compete in the United States, because their Caltex joint venture in the Eastern Hemisphere produces a substantial portion of the net profit of each of these U.S. companies.

We have responded to Senator Tunney by letter in which the facts about Caltex are set out. I wish to assure the subcommittee today, without fear of contradiction, that competition between Socal and Texaco in the U.S. is as intense and vigorous as it is with every other company. For instance, we have made 302 competitive bids against Texaco bids on the same OCS tracts. We submitted the highest bid 35 times, while Texaco submitted the highest bid 83 times. Not once have we bid jointly with Texaco on an OCS tract.

Joint ventures and farmouts are indispensable tools in exploration and production. Because of the great number of operators now engaged in exploration and production and the keen competition for leases, it rarely occurs that all of the acreage in a new prospect is owned by a single operator. Given more than one lease owner in an area, some kind of joint operating arrangement is almost inevitable. Indeed, in states that have laws designed to conserve natural resources, operations without joint venture agreements would not be feasible.

Joint operating arrangements help spread the high risk and great expense of exploring for and drilling and producing in difficult operating areas such as the Gulf of Mexico, Alaska and the North Sea. In such areas, even the largest major companies are forced to seek partners who are willing to share the costs and risks.

Joint venture agreements, like joint bids, enable the non-major companies to take part in exploration and production in high risk and expense areas.

CONCLUSION

Finally, let me sum up exactly what is at stake.

The advocates of legislation to break up the largest integrated companies must accept an awesome responsibility for what they propose to do.

The United States has the largest Gross National Product of any nation in the world. Yet at the same time, we have a dangerous and growing energy gap—a dependence on foreign sources for our most essential energy needs.

Because of unwise policy in the past, which denied sufficient financial incentives to produce greater domestic energy, this nation faces the most critical economic challenge since World War II.

We are painfully and slowly emerging from the deepest recession of the post-war years. Unemployment is a serious national problem. Yet any disruption of energy supplies can cause widespread unemployment—as we learned during the Middle East oil embargo.

It is against this background that the advocates of divestiture must weigh their action. Can it be seriously suggested that at this critical time, America should systematically dismantle, disrupt and cripple the companies best equipped to contribute the most to solving our energy problems?

The whole premise on which this legislation is based is false. Although a suit is currently pending against eight of the largest companies, charging violation of the Federal Trade Commission Act, twice in the past year Administrative Law Judges in charge of the case have suggested that the Commission reconsider taking the case further.

The charge of monopoly remains unproven because it is simply not true. While it is often claimed that unreasonable concentration exists in the oil industry, the facts show that petroleum is one of the least concentrated of the major U.S. industries.

If no monopoly exists, dismemberment of the industry is not only an injustice, it will be a tragic misapplication of government power that will hurt America's economy and make it more difficult for this country to achieve greater energy self-sufficiency. Breaking up integrated companies won't reduce the price of gasoline or any other product. Instead, it will set in motion a chain of events that will make America even *more* dependent on high priced foreign oil.

If the largest integrated companies are torn apart, I am convinced they will have to be put back together again in essentially the same form, if America hopes to have enough energy for a prosperous and expanding economy.

Those who propose divestiture must bear the burden of responsibility for the disruption that will follow in the wake of divorcement: for the energy shortages that will result, for the unemployment that will be caused by a shortage of oil and gas.

Advocates of divorcement will have to answer to the millions of people who have invested their savings in the petroleum industry.

More than 2.3 million individual citizens own stock in the six largest integrated companies which would be broken up by this legislation. Of these, 46 per cent are retired people, living on pensions and dividends. And another 11 million people have indirect ownership through the stock held by

— 91 colleges and universities

— pension plans for state and municipal employees

— retirement programs for unions and companies

— nearly 200 mutual insurance companies

— almost 1,000 charitable and educational foundations

They have invested their savings and their assets in these companies because the U.S. petroleum industry is the world's most efficient, with a long and proven record of producing the energy our country needs at the least possible cost to consumers.

Tearing this industry apart through divestiture will carve this industry into four segments, each untested in financial markets and each scrambling to reorganize into a competitive company in an industry where growth will have become an economic crime—punishable by dismemberment.

It is impossible to forecast anything but catastrophic consequences if this legislation is passed.

It does not make sense to order divorcement simply to satisfy some abstract and unproved theory of competition.

The facts demonstrate that integration *does not* unreasonably restrain competition. On the contrary, it intensifies competition. Divorcement will produce less and more expensive energy.

I hope the views I have presented will be helpful to you, and I respectfully urge you to consider them and reject S.2387 and similar proposals before you.

EXHIBIT 1

Exploration for and production of crude oil and natural gas are the first steps in the intricate and sophisticated processes that culminate in the availability of finished petroleum products for public consumption. Whether the public demand for petroleum is adequately met depends directly and inevitably upon the success of the exploratory and producing effort around the world.

Exploration begins long before the drilling of the first well. In any company, geophysicists and geologists assemble and analyze data, old and new, in search of likely prospects. In the course of these studies an integrated company frequently uses information and techniques that have been pioneered and perfected by the research and development arm of the company. Similarly, they seek the assistance of the company's geophysics division, which operates costly and complex computer systems designed to interpret raw geophysical data rapidly and accurately.

If a likely prospect is encountered as a result of these studies, oil and gas leases are obtained through the efforts of the land staff of the company. If the prospect includes lands owned by the Federal government or a state government, it is usually necessary to bid for leases at a public sale. Since public bidding most often involves substantial cash expenditures for bonuses and rentals, the talents and expertise of still another unit of the company—the finance department—are utilized. And even more importantly, the money to make the bid has to be found within the total resources of the integrated company or borrowed from financial institutions or others based upon the overall financial standing and financial integrity of the entire company.

When leases have been obtained, the prospect is ready for more detailed scrutiny by the geologists and geophysicists to pinpoint a location for the initial exploratory well. Once again the assistance of the research division and of the geophysics division of the integrated company may be sought to provide and to process additional information. When a location has been selected, the company's land staff may be called upon to secure various permits and rights-of-way that are required from government agencies, landowners and others.

When drilling begins on the prospect, a new group of experts and specialists enter the picture—the engineers of the company who are responsible for drilling and operating oil and gas wells. From this point forward there is a close coordination of the efforts of exploration geologists, development geologists, petroleum engineers, construction engineers and others.

After the drilling of a successful offshore exploratory well, decisions must be made about where and when development wells are to be drilled. Drilling platforms and producing facilities must be designed, built and installed. Facilities to handle production must be provided. These include gathering lines for oil and water; natural gas pipeline connections; treating facilities to clean up oil production to meet pipeline specifications; natural gas processing facilities to separate natural gasoline, butane and propane; and facilities to remove contaminants such as carbon dioxide and hydrogen sulfide which are sometimes needed to make the gas marketable. In addition, other necessities must be added for sales contracts for crude oil, natural gas and plant products.

At every step innumerable permits must be obtained from Federal, state and local regulatory agencies. Financial questions, land and lease questions, legal questions and other questions must be considered and dealt with. Many of the specialized talents of the integrated organization are utilized.

From the foregoing it is readily apparent that exploration and production in an integrated company are not self-sufficient activities conducted independently of other departments in the company. On the contrary, these other departments contribute immeasurably to the efficiency and effectiveness of the exploratory and producing efforts. Without such contributions the costs of exploring and producing would be much quieter. The use of complex and expensive technology would also fall off, and exploration and production would decline, with a consequent loss in supplies of oil and gas. Thus, divestiture would not only fail to *increase* petroleum supplies, but would actually decrease them.

EXHIBIT 2

NET PRODUCTION OF CRUDE OIL AND NATURAL GAS LIQUIDS AND PRODUCTION SHARES OF INDIVIDUAL COMPANIES—1974

Rank	Company Name	Crude Oil and NGLs Barrels per Day (1,000)	% of U.S.
1	Exxon	890.0	(8.5)
2	Texaco	705.0	(6.7)
3	Shell	586.0	(5.6)
4	Standard of Indiana	539.0	(5.2)
5	Socal	413.1	(4.0)
6	Gulf	400.9	(3.8)
7	ARCO	383.1	(3.7)
8	Mobil	363.0	(3.5)
9	Getty (T)	269.7	(2.6)
10	Union	269.6	(2.6)
11	Phillips	255.7	(2.5)
12	Sun	219.3	(2.1)
13	Cities Service	213.0	(2.0)
14	Conoco	188.0	(1.8)
15	Marathon	174.0	(1.7)
16	Amerada Hess	98.8	(1.0)
17	Tenneco	78.8	(0.75)
18	Louisana Land & Exploration	77.2 (0)	(0.74)
19	Superior Oil	68.3 (B)	(0.65)
20	Pennzoil	54.3 (S)	(0.52)
			(% of U.S.)
	Total for Top 4 Companies	2,720	(26.0)
	Total for Top 8 Companies	4,280	(41.7)
	Total for Top 20 Companies	6,247	(59.8)
	U.S. Total	10,453	(100.0)

EXHIBIT 2 (CONT'D)

NET PRODUCTION OF NATURAL GAS AND PRODUCTION SHARES OF INDIVIDUAL COMPANIES—1974

Rank	Company Name	Natural Gas Cu. Ft. per Day (Millions)	Natural Gas % of U.S.
1	Exxon	5,312.0 (N)	(9.3)
2	Texaco	4,097.0 (N)	(7.2)
3	Standard of Indiana	2,893.0	(5.1)
4	Mobil	2,281.0	(4.0)
5	Gulf	2,183.0	(3.8)
6	Shell	2,050.0	(3.6)
7	ARCO	1,739.0 (N)	(3.0)
8	Phillips	1,433.0	(2.5)
9	Union	1,384.4	(2.4)
10	Sun	1,307.0 (N)	(2.3)
11	Socal	1,289.0	(2.3)
12	Getty (T)	1,185.7	(2.1)
13	Cities Service	1,115.0	(2.0)
14	Tenneco	1,040.0	(1.8)
15	Conoco	897.0	(1.6)
16	Superior Oil	890.5 (B, N)	(1.6)
17	Pennzoil	748.0	(1.3)
18	El Paso	625.0	(1.1)
19	Marathon	408.4	(0.71)
20	Amerada Hess	357.1	(0.62)
			(% of U.S.)
	Total for Top 4 Companies	14,583	(25.5)
	Total for Top 8 Companies	21,988	(38.4)
	Total for Top 20 Companies	33,235	(58.1)
	U.S. Total	57,212	(100.0)

FOOTNOTES:

(B) Gross.
(N) Marketed Sales.
(O) Includes Working Interest and Royalty Interest Properties.
(S) Includes 100% for POGO Interest but does not include any amounts attributable to PLATO.
(T) Includes Skelly and Mission Corp.
(U) Estimated, based on 1974 total sales and ratio of 1973 "owned" to "contracted and purchased."
NGL Natural Gas Liquids

EXHIBIT 3

REASONS WHY THERE IS NO INCENTIVE FOR SHUTTING-IN OIL PRODUCTION

The nature of oil reservoirs, oil wells and oil fields limits production rates and rules out most incentives to hold back oil production in anticipation of a price increase. Most importantly, production rates cannot be jumped to recover, in short order, oil delayed by a shut-in of a year or so.

After an initial short period at a stable rate, the production from a reservoir begins to decline. For most reservoirs this decline will be directly related to the reserves left in the ground. As oil is produced, the oil remaining in the reservoir occupies less space. Typically either gas or water displaces the oil, but in either case, there is less remaining oil-filled rock through which oil may flow to the well. All wells in a reservoir will not react exactly alike, but on the average, production will decline as reserves decline.

Most wells in the U.S. have reached this stage and production typically declines 10 per cent per year. Thus, a well that would have a production of 10,000 barrels in one year, would produce only 9,000 barrels the next year, 8,100 barrels in the third year, and so on. The production rate continues to diminish until, at the end of the well's life, the income no longer meets expenses—20, 30, 40 or even more years after discovery.

If a well is shut-in and subsequently returned to production it will usually stabilize to the same production and decline rate, except for the time lag, as if shut-in had not occurred. If the example well were shut-in for a year, it would produce only 10,000 barrels the next year rather than the 9,000 barrels after being continuously produced. Only 1,000 barrels would have been regained of the 10,000 barrels deferred. Unlike wheat in an elevator, which might be held for a year and sold with the new crop the following year, deferred oil production cannot be fully recovered until the end of the life of the well. Extending the example for a few more years to compare shut-in and continuous operation, the annual production would be, assuming shut-in sometime during the life of a well after decline has begun:

Production in Barrels

Years After Start of Shut-In	Continues Producing	Shuts-In For A Year	Loss of Oil Due To Shut-In	
			Annual	Cumulative
1	10,000	0	10,000	10,000
2	9,000	10,000	–1,000	9,000
3	8,100	9,000	–900	8,100
4	7,290	8,100	–810	7,290
5	6,560	7,290	–730	6,560
6	5,900	6,560	–660	5,900
7	5,310	5,900	–590	5,310

Now let's compare the financial effects:

. . . . .assuming the example well could get $5 per barrel for oil in the first year and expects $10 per barrel during the second and subsequent years and direct operating costs are $1.50 per barrel.

EXHIBIT 3 (Cont'd)

Revenue From Sale of Crude Less Direct Operating Costs

Year	Continues Producing	Shuts-In For A Year	Increased Net Revenue Due To Shut-In	
			Annual	Cumulative
1	$85,000	0	$-35,000	$-35,000
2	76,500	$85,000	+8,500	-26,500
3	68,900	76,500	+7,600	-18,900
4	62,000	68,900	+6,900	-12,000
5	55,800	62,000	+6,200	-5,800
6	50,200	55,800	+5,600	-200
7	45,200	50,200	+5,000	+4,800

In this case the operator would not regain the dollar loss he incurred by shutting-in the first year until sometime in the seventh year. Not then if some direct costs continued even when the well was shut-in. Assuming costs and the $10 price held constant for the remaining life of the well, *eventually* the operator would gain $50,000 (10,000 x $5). But using any of the current proposals for phasing out price controls and assuming a reasonable time value of money, the loss likely would never be regained.

This basic principle applies to all producers—small, large, independent and integrated, and to both oil and gas wells. It also applies to cutback as well as shut-in production.

This example demonstrates one of the reasons that operators want to produce oil wells at the maximum rate. *Stockholders want dividends every year* as well as reinvestment of earnings to maintain the value of their shares. They would not be willing to wait seven years for their money—more likely they would *fire the manager* in the interim. The same problems apply to *financing an election campaign*. Cash received seven years in the future will be of little use for a campaign next year.

The following factors may also encourage continued production:

1. Drainage of reserves to nearby wells which continue producing could result in a net loss of oil.

2. Mechanical problems with the well could result from a prolonged shut-in period.

3. Royalty owners often demand a continuing income; thus, some leases become void if production is interrupted.

4. The effectiveness of assisted recovery programs could be permanently reduced from being shut-in for an extended period.

5. The depletion allowance for small producers will be progressively reduced.

Economic as well as physical and legal considerations all indicate that shutting-in or cutting-back production is unattractive in the vast majority of cases.

EXHIBIT 4

UPDATE OF TABLE IV OF AAPG BACKGROUND PAPER #2, 1975 RESERVES CLASSIFICATION FOR NEW FIELD WILDCAT SIGNIFICANT DISCOVERIES, 1969-1974 (Million Oil Equivalent Barrels*)

	Industry		Independents			Majors (16)		
Size Classification	No. Wells	Reserves	No. Wells	Reserves	(%)	No. Wells	Reserves	(%)
Oil Discoveries								
A	12	1,200	2	200		10	1,000	
B	7	259	0	—		7	259	
C	15	255	9	153		6	102	
D	261	783	201	603		60	180	
Subtotal	295	2,497	212	956	(38.3)	83	1,541	(61.7)
Gas Discoveries								
A	6	600	2	200		4	400	
B	10	375	7	263		3	112	
C	47	822	28	490		19	332	
D	357	1,175	288	950		68	225	
Subtotal	419	2,972	325	1,903	(64.0)	94	1,069	(36.0)
TOTAL	714	5,469	537	2,859	(52.3)	177	2,610	(47.7)

*Gas converted to oil equivalent barrel on basis of 6,000 cubic feet to one barrel of oil.

NOTE: These reserve figures do not portray actual field sizes. For example, the reserve size attributed to any major discovery (i.e., Purdhoe Bay) is only 100 million barrels.

(Source: AAPG, Strategic Committee on Public Affairs in cooperation with the committee on Statistics on Drilling; see also references Table I, November, 1975.)

7

Vertical Divestiture: Refining

TESTIMONY OF WALTER R. PEIRSON, PRESIDENT, AMOCO OIL COMPANY, BEFORE THE SUBCOMMITTEE ON ANTI-TRUST AND MONOPOLY, COMMITTEE ON THE JUDICIARY, UNITED STATES SENATE — SEPTEMBER 26, 1975

Summary: The proposed divestiture legislation erroneously assumes that the refining industry is concentrated. It is not. Twenty-six other basic industries are much more concentrated. Vertical integration runs common to industry generally, one of its efficiencies being elimination of large inventories because of security of supply. Long-term contracts would replace vertical integration for that purpose but with less efficiency and higher cost. Divestiture would produce an economic loss of $18 billion in domestic assets for the holders of oil company securities.

Walter R. Peirson speaks . . .

My name is Walter R. Peirson, and I am President of Amoco Oil Company—the domestic refining, marketing and transportation subsidiary of Standard Oil Company (Indiana). I am grateful for the opportunity to appear before you today to state my Company's position on S. 2387 and to present the basis for our belief that this proposed legislation is based on an inaccurate view of the petroleum industry; is arbitrary and irrational in its approach to the subject of vertical integration; and will, if passed, impose a heavy toll on the consumer while producing none of the alleged procompetitive effects which its proponents have predicted. In light of other testimony already presented to this panel, I will address myself principally to the impact of vertical integration on the refining of petroleum.

We believe that this proposed legislation is arbitrary and is based on a misapprehension of the structure of the petroleum industry for the following reasons. The proposed bill erroneously assumes:

First, that the industry is highly concentrated;

Second, that the companies subject to its provisions are materially different from other significant refiners in their degree of vertical integration; and

Third, that the degree of existing integration has impeded successful entry into the refinery market and growth. These contentions simply are not supported by the facts.

Despite repeated assertions to the contrary during the proceedings of this Subcommittee, the petroleum refining industry is not highly concentrated under any meaningful measure of concentration. As shown in Exhibit A, attached to this statement, twenty-six basic industries have higher concentration ratios than petroleum refining and several, including primary aluminum, basic sheet glass, the automotive industry, steam engines and turbines, primary copper and the rubber industry are two to three times more concentrated than refining.

There are presently more than 130 refiners operating in the United States. The largest refiner has an 8 percent share of U.S. refining capacity. It perverts economic theory to assert that 130 refiners with such modest market shares constitute a concentrated industry.

To avoid the obvious conclusion compelled by these data, both proponents and sponsors of this legislation have taken refuge in the alleged synergism of this degree of concentration when combined with vertical integration. A sponsor of this legislation has observed, that ". . .the domestic oil industry is essentially noncompetitve because of a unique combination of extensive vertical integration and concentrated economic power which enables a relatively few companies to maintain excessive control over the industry."

This position reveals a misunderstanding not only of the industry, but of the nature of vertical integration itself.

Exhibit B lists the 19 refiners whose operating capacity would classify them as a major refiner under this proposed statute. Not all of these companies are active at every vertical level of operation.

Exhibit C lists the 45 next largest refiners in the United States as measured by their operating crude running capacity as of January, 1975. These companies range in national market share from 1.27% for American Petrofina to .13% for Southland, Crystal Oil and Witco Chemical. This obviously represents a full spectrum of companies in the industry. It includes companies of varying size, financial resources, and geographic dispersion. It includes companies that have been viewed traditionally as "oil companies," whose primary capital commitment and income are related to the energy business. It also includes companies such as Union Pacific and Studebaker Worthington that have entered refining through diversification.

It can be seen that of these 45 refiners, 33 produce crude, 32 own crude oil pipelines, 25 own refined products pipelines, and 35 are branded gasoline retailers. Contrary to the assumption that vertical integration is limited to a few dominant majors, it is clear that some degree of vertical integration is the rule rather than the exception for many refiners. Indeed, 19 of these 45 refiners are "totally integrated" in the sense that they perform some activity at each vertical level of the industry.

The proponents of this legislation are incorrect not only in their views of the prevalence of vertical integration, but also as to its impact. If vertical integration conferred the market power ascribed to it by certain witnesses before this Committee, one would expect an ever-increasing percentage of total refining capacity controlled by large refiners, a dramatic decrease in the number and total refining capacity of smaller refiners, and a marginal "independent" segment composed of anemic and ineffective competitors, sustained only by scraps thrown by the dominant majors.

Quite the contrary is true. There have been approximately 50 new entrants in refining since 1950 which are still operating. Only 6 of these companies entered with refining capacity in excess of 16,500 barrels per day. Eight entered with refineries rated between 10,000 and 16,500 barrels per day, and most of the remaining entrants started with less than 5,000 barrels per day.

Starting small apparently has not inhibited growth. As Exhibit D indicates, the capacity of 22 refining companies grew to more than 50,000 barrels per day between 1951 and 1975. Eight of these firms, including the first and third largest in 1975, entered the refining business during this period. Two of these refiners, Amerada Hess and Coastal States, have been such successful competitors that their dramatic growth qualifies them for divestiture under this bill. The aggregate capacity in 1951 of the other 14, or their predecessors, was only 239,200 barrels per day. The aggregate capacity of the 22 in 1975 was 3,181,440 barrels per day. This is a 1233% increase compared with the 134% increase for all U.S. refineries.

Thus, the barriers to entry allegedly raised by vertical integration have not impeded growth or prevented successful entry at widely different levels of operation.

There has been testimony before this Subcommittee that vertical integration has enabled major refiners to shift profits to the production level and to "squeeze" the margins of independent refiners. The statistics on the growth which I have just reviewed refute that assertion. Prior to the Oil Embargo of 1973, there was adequate competitively priced crude for the smaller refiner, and during the shortage period the FEA made crude available to all parties.

Moreover, integrated firms have neither power nor incentive to shift profits from one phase of their operations to another. The lack of concentration at the production level precludes the power to limit supply. In addition, there would be no incentive for integrated companies to sell refined products for less than competition would allow. It appears that all but one of the companies subject to this bill are net purchasers of crude. To forego earnings in refining, even with increased earnings in production, would result in less total profit, not more.

In its Report submitted to the Congress on the impact of decontrol of petroleum prices, the Federal Trade Commission concluded that there was no serious risk that such price decontrol would lead to a debilitating price squeeze on nonintegrated refiners. While the staff concluded that some downward pressure on refining margins was probable, this result would not be undesirable if excess capacity conditions persisted at that time. The Report also concluded that while the elimination of existing subsidies and possible downward pressures on refining margins might hurt the least efficient refiners, it is important to distinguish between possible injury to a specific competitor and generalized injury to the vigor of competition.

Throughout the proceedings of this Subcommittee, the record reveals a pervasive, yet unstated belief that vertical integration in the petroleum industry is the product of a conspiratorial scheme by major oil companies to maximize industry control. Such a belief is unfounded.

I would like to illustrate how a decision to integrate is reached by describing the experience of my Company in the construction of its two most recently completed refineries. The result of one project was a refinery which some might regard as demonstrating a high degree of "vertical integration" since Amoco constructed, in addition to the refinery, an extensive system of gathering lines to bring crude oil to the refinery and pipelines to deliver refined products to market. The other refinery is virtually completely "nonintegrated."

In the late 1940's, a major new crude oil reservoir was discovered in the Williston Basin in North Dakota. The commercial development of this significant new discovery was demonstrably in the interest of the public. However, this project was stymied by one of the most significant obstacles to production and utilization of crude: the crude was land-locked; the area had had no prior significant production and no pipelines existed to transport the crude to a refinery. In this area, the most economic method to transport crude oil in volume is by pipeline.

Amoco itself had only a very small producing interest in Williston Basin crude. Its objective was primarily that of a refiner: to secure a substantial, new supply of crude located near its north midwestern markets.

This sparsely populated area, however, could not support a refinery larger than 15,000 barrels a day. Given the cost of building a refinery even in the 1950's it would have entailed a large investment to process a very minor portion of available crude.

The solution was to construct a larger, more efficient refinery which was in scale with forecast crude production. This refinery, however, would require a crude oil gathering system and a refined products pipeline to transport products to markets beyond the Dakotas. There were obviously only two ways to construct these pipelines: Amoco could do it, or a third party could do it.

Several considerations compelled the conclusion that the only realistic alternative was for Amoco to do it. If we were to capitalize on the availability of this new-found crude oil it was important to construct the refinery as quickly as possible. There simply was not time for a prolonged campaign to stimulate interest in a project with a capital commitment of this magnitude. Morecver, the lines had to be built in tandem with the refinery. We could not risk a refinery without crude to process or an operating refinery with no outlet for its products. Accordingly, logic and economics compelled Amoco to build this highly integrated pipeline and refinery system at Mandan, North Dakota.

Amoco's Mandan refinery began operation in 1955 and provides a stable and secure source of refined products for the Dakotas, and its pipelines transport refined products to the Twin Cities. In this instance, Amoco integrated because it saw no other practical alternative.

This instance of integration can be contrasted with our Yorktown, Virginia refinery which is of approximately the same size and age. The Yorktown refinery started operation in 1956 to process imported crude oil almost exclusively. Existing means of transportation gave Amoco adequate access to the northeastern markets this refinery was intended to serve. Prior to the construction of the Yorktown facility, Amoco had no significant East Coast refinery. Its construction promoted competition in these markets, and the consumer received the benefits of low-cost raw material and efficient operations.

In each of these instances, Amoco's decisions were made on the basis of the economics of each situation; and the "integration" of the Mandan project and the "nonintegration" of the Yorktown project both fostered the one basic objective which this proposed bill overlooks—the most efficient and economical production of refined products for consumption by the American people.

If Amoco were prohibited by law from building the crude and refined products pipelines for the Mandan Refinery, the results would inevitably have been delay and increased transportation expense. This legislation will eliminate the flexibility to evaluate on a decision-by-decision basis the economics of growth and require multiple middlemen where the consumer would clearly benefit through the economics of integration.

Vertical integration is a universal business phenomenon. It occurs in nonconcentrated as well as concentrated industries and in large and small firms. Exhibit E lists 16 industries and describes the extent of vertical integration in each. Virtually each of these industries is as integrated as refining, and many are considerably more integrated. In the radio and television industry, for example, a firm typically produces the programs which originate from and are broadcast on a national network of affiliated television and radio stations and received through antennae, radios and television sets manufactured by other affiliates. A thoughtful review of this exhibit compels the conclusion that some concern other than vertical integration has motivated this legislation which penalizes the oil industry for an economic structure common to industry generally.

The historic focus of our antitrust laws has been the elimination of anticompetitive conduct. The legislation is nothing less than a revision of the nation's antitrust policy through the misapplication of an economic theory of concentration to an unconcentrated industry. It is noteworthy that this theory has been questioned, and, in some cases, repudiated, by leading economists. Let us not assume that the impact of this decision will be limited to the oil industry. The atmosphere of crisis which has induced this legislation may well be repeated, and other industries may be faced with similar proposals as a result of unexpected shortages or adverse economic developments. The Congress should address the broad issue of the impact of deconcentration on our entire industrial base.

Let us assess the costs of this policy before plunging headlong into the dismemberment of the world's most efficient petroleum industry. The first is the cost of divorcement itself. Divorcement by sale of assets is improbable. During 1974, 6 billion dollars of new equity funds were generated in the U.S. for all corporate purposes. The net worth of the 19 largest oil refiners is approximately 72.4 billion dollars. An attempt to generate new equity on that scale would cause a capital crisis of historic proportions.

Senator Gary Hart has suggested that an outright stock distribution to existing shareholders would easily accomplish divorcement. His characterization of this alternative as a "paper transfer" totally overlooks several fundamental obstacles.

What will be done with corporate debt? Standard Oil Company (Indiana) has 1.4 billion dollars in long-term debt held by more than 27,000 individual bondholders. Many of these bondholders are corporate and individual trustees for pension and retirement funds whose beneficiaries are hundreds of thousands of average citizens for whom this investment is a significant part of life savings. In 1974, Standard paid $112 million in interest to these bondholders. This debt was acquired on the basis of Standard's assets and historical earnings stability. Each of the indentures pursuant to which these debentures were issued prohibits unilateral assignment of the repayment obligation. If Standard Oil were to be broken up into production, transportation, refining and marketing components, each of these new companies would be expected to bear some appropriate part of this debt. If the bondholders refused to accept a restructuring of Standard's obligation, the newly created companies would have to raise sufficient new capital to retire whatever portion of Standard's total has been allocated to it.

However, the 19 refiners have total outstanding long-term debt of approximately 19.1 billion dollars. Refinancing of this debt by the spun-off companies may not be possible. Even if refinancing were possible, it is our projection that the spun-off entities would have increased annual interest expense of more than 500 million dollars.

One of the obvious efficiencies of vertical integration is the elimination of the unnecessarily large inventories because of security of supply. Initially, the refining and marketing segments of the business would find it necessary to increase inventories to have sufficient raw material, intermediate stock, and finished product for contingencies and to insure low-cost purchase through adequate storage flexibility. The cost of unnecessary, increased inventories would divert capital much needed for essential supply development and expansion of existing facilities. When these costs are added to the inherent inefficiency of companies denied any of the flexibilities and cost savings of even partial integration, the cost of refined products *must* be higher.

This Subcommittee has given no thought to the impact of divorcement on the millions of individual stockholders who own these companies which you would dismember. These are average people, many of them retired, who selected this broad based stock because of its historic reliability and security. This legislation will leave them with shares in several new companies with no earnings and stock valuation history. The risk premium reflected by the increased annual interest expense mentioned above will impact on the entire asset base of these companies, not just the debt. We anticipate that bondholders and stockholders will suffer an economic loss of approximately $18 billion on domestic assets. This figure does not include the truly staggering cost if foreign integrated assets and debt of multinational companies are divested.

Analogies to the rapidly appreciated value of shares of the companies spun off from the Standard Oil trust are inappropriate. Each of these companies was at least partially vertically integrated and began independent operations during the period of greatest economic growth in America's history. I am not saying that the entities which would be created by this legislation must or will fail. However, your constituents—and I am certain that each of you has many constituents who are shareholders in these companies—deserve a more careful scrutiny of the impact of this legislation on their investment than they have been afforded.

Finally, we must ask the ultimate question: Will these social and economic costs be offset by pro-competitive effects of the creation of a myriad of purchasers and sellers? We believe not. Eventually, the prospect of raw material cost advantages and security of supply will lead to long-term contracts at each level of the industry. We project that a contractual system would gradually evolve which would essentially duplicate the vertical integration which exists today, but with less efficiency and a higher level of cost.

We believe that this bill will fail to enhance competition, and fail at a terrible cost. We believe further that the manner in which it is being rushed through Committee with no apparent study of the cost to the American public is irresponsible. Hopefully, the day will arrive when the citizens of this country will awaken to the fact that they have been conned into the trap of wishfully thinking that punitive legislative proposals of this type, which would have the effect of destroying, regulating and burdening one of the primary assets of this country, will somehow make the energy problem disappear.

This legislation is being considered at the very time that both the OPEC nations and consuming countries are moving to reduce the ownership and market share of U.S. companies. This bill will eliminate the ability to meet increased competition from growing integrated private and publicly held foreign companies with fateful consequences for the nation's position in international trade.

The structure of the petroleum industry is competitive. It produces more energy at lower cost to the consumer than any in the world. The radical surgery which this legislation contemplates cannot be undone if our predictions become reality and the price of this hasty and ill-conceived measure will befall the American people at the time in our history when the delivery of adequate energy is a critical national problem. We urge you to reject S. 2387.

EXHIBIT A

1970 CONCENTRATION RATIOS FOR REPRESENTATIVE INDUSTRIES

S.I.C Code		4-Firm Ratio	8-Firm Ratio
3741	Locomotives and parts	97	99
3334	Primary aluminum*	96	100
3211	Basic sheet glass**	94	98
3641	Electric lamps	92	97
3711	Automotive industry	91	97
3672	Cathode ray picture tubes	88	97
2073	Chewing gum	85	97
3633	Household laundry equipment	83	97
3572	Typewriters**	81	99
2111	Cigarettes**	81	100
3511	Steam engines and turbines	77	88
3331	Primary copper	75	98
3011	Rubber industry	72	89
3411	Metal container industry	72	83
2841	Soap and detergents	70	79
3721	Aircraft and parts	65	87
2822	Synthetic rubber	64	85
2284	Thread mills	64	83
3221	Glass containers	57	78
3871	Watches and clocks	56	70
3562	Ball and roller bearings	54	74
3621	Motors and generators	50	62
3651	Radio & TV Receiving sets	48	67
3312	Blast furnaces and steel mills	47	65
2082	Beer and malt liquors	46	64
	All manufacturing	40	NA
3522	Farm machinery & equipment	40	51
2911	Petroleum refining	33	57
2211	Broad woven cotton mills	33	50
3552	Textile machinery	33	49
2041	Flour mills	30	46
2051	Bread and related products	29	39
3141	Shoes, except rubber	28	36
3241	Cement	27	48
2834	Pharmaceutical preparations	26	43
3541	Metal-cutting machine tools	24	37
2851	Paints and allied products	22	34
2026	Fluid milk**	22	30
2256	Knit fabric mills	17	28
2311	Men's & boys' suits & coats**	17	27
2711	Newspapers	16	24
2421	Sawmills & planing mills	16	20
3494	Valves and pipe fittings	14	23

2511	Wood furniture, not upholstered	14	23
3251	Brick and structural tile**	14	22
2086	Bottled & canned soft drinks	13	20
3451	Screw machine products**	6	9

*1963 figures
**1967 figures

Source: U.S. Department of Commerce, Annual Survey of Manufacturers, 1970.

EXHIBIT B

"MAJOR" REFINERS SUBJECT TO S. 2387

	Capacity Bureau of Mines 1/1/75*
Amerada Hess Corp.**	730,000
Ashland Oil, Inc.	362,943
Atlantic Richfield	724,600
Cities Service Co.	268,000
Coastal States Gas Corp.	212,982
Continental Oil Co.	377,000
Exxon Corp.	1,243,000
Getty Oil Co.	218,731
Gulf Oil Corp.	866,400
Marathon Oil Co.	324,000
Mobil Oil Corp.	903,500
Phillips Petroleum Co.	408,000
Shell Oil Co.	1,150,000
Standard Oil Company of California	952,000
Standard Oil Company (Indiana)	1,115,000
Standard Oil Company (Ohio)	431,000
Sun Oil Co.	484,000
Texaco Inc.	1,073,000
Union Oil Company of California	496,000

*Barrels per day.
**Includes the capacity of the Virgin Islands refinery on 12/1/74 as reported by the FEA.

EXHIBIT C

VERTICAL INTEGRATION OF CERTAIN REFINING COMPANIES[1]

	Operating Refining Capacity 1/1/75					
	B/D	% of U.S. Total	Produces Crude Oil	Owns Crude Oil Pipelines	Owns Refined Products Pipelines	Markets Branded Gasoline
American Petrofina	200,000	1.27	x	x	x	x
Kerr McGee Oil Industries	166,000	1.06	x	x	x	x
Commonwealth Oil Rfg.	161,000	1.02			x	x
Union Pacific (Champlin)	152,200	.97	x	x	x	x
Murphy Oil Corp.	137,500	.87	x	x	x	x
Koch Industries	109,800	.70	x	x	x	x
Clark Oil & Refining	108,000	.69	x	x	x	x
Tenneco Oil Co.	103,000	.65	x	x	x	x
Crown Central Petroleum	100,000	.64	x	x	x	x
Oil Shale Corp.	87,000	.55		x	x	x
Charter Co.	85,900	.55	x	x	x	x
Agway, Inc. (Texas City Rfg.)	74,500	.47	x			x
Farmland Industries (CRA)	73,838	.47	x	x		x
Tesoro Petroleum Co.	64,000	.41	x	x		x
Pennzoil Co.	62,600	.40	x	x	x	x
Husky Oil Co.	59,000	.38	x	x	x	x
Apco Oil Co.	58,670	.37	x	x	x	x
United Refining	58,000	.37	x	x	x	x
National Cooperative Rfg. Assn.	54,150	.34	x	x	x	x
Placid Oil Co. (Toro Petr. Co.) Hunt Oil Co.	36,000) 18,000)	.34	x	x	x	x

Studebaker Worthington (Pasco)	48,600	.31	x	x	x	x
Diamond Shamrock	48,300	.31	x	x	x	x
Powerine Oil Co.	46,000	.29	x	x		x
Earth Resources (Delta Rfg.)	43,900	.28				x
Total Petroleum	42,182	.27	x	x	x	x
Farmers Union Central Exchange	41,650	.26	x	x	x	x
Hawaiian Independent Rfg.	40,389	.26				
Pride Refining	36,500	.23				x
Howell Hydrocarbons	37,000)					
Quintana Howell Joint Venture	30,000)	.21	x			
Southern Union Gas Co. (Famariss Oil & Rfg. Co.)	33,000	.21	x			x
Esmark (Vickers)	32,500	.21	x	x	x	x
Holly Corp. (Navajo Rfg.)	29,900	.19	x	x	x	
Rock Island Refining Co.	29,500	.19		x		x
Edgington Oil Co.	29,500	.19	x	x		
Texas Eastern Transmission	29,500	.19	x	x	x	
Gasland, Inc. (Good Hope Rfg.)	29,450	.19				
San Joaquin Oil Co.	27,000	.17				
OKC Corp	25,000	.16		x		
Quaker State Oil Rfg. Co.	24,500	.16	x	x	x	x
Little America Rfg. Co.	24,300	.15				x
Reserve Oil & Gas Co. (Mohawk)	22,100	.14	x	x		x
First General Resources (The Refinery Corp.)	21,500	.14				
VGS Corp. (Southland)	21,000	.13				
Crystal Oil Co.	20,300	.13	x			x
Witco Chemical Co.	20,000	.13	x	x		x

[1]These are the 45 largest refining companies, other than those that would be affected by S. 2387, listed in order of their size as measured by their operating crude running capacity on 1/1/75.

EXHIBIT D

REFINING COMPANIES WHOSE OPERATING CRUDE OIL DISTILLATION CAPACITIES GREW TO MORE THAN 50,000 B/D BETWEEN 1/1/51 AND 1/1/75

	Operating Refining Capacity[1]	
	on 1/1/51	on 1/1/75
Amerada Hess Corp.	- 0 -	730,000
Marathon Oil Co.	31,000	324,000
Coastal States Gas Corp.	- 0 -	212,982
American Petrofina	19,800	200,000
Kerr McGee	7,500	166,000
Commonwealth Oil Rfg.	- 0 -	161,000
Union Pacific (Champlin)	20,200	152,000
Murphy Oil Corp.	- 0 -	137,000
Koch Industries	- 0 -	109,800
Clark Oil & Rfg. Co.	26,000	108,000
Tenneco, Inc.	16,000	103,000
Crown Central	32,500	100,000
Toscopetro (The Oil Shale Corp.)	- 0 -	87,000
Charter Co.	10,000	85,900
Agway, Inc.	- 0 -	74,500
Farmland Industries	27,200	73,838
Tesoro Petroleum Co.	- 0 -	64,000
Pennzoil Co.	8,500	62,600
Apco Oil Corp.	10,000	58,670
Husky Oil Co.	5,000	59,000
United Rfg.	5,500	58,000
National Coop. Rfg. Assn.	20,000	54,150
Total of 22 Companies	239,200	3,181,440

[1]These capacities are as reported by the Bureau of Mines, except that they include the capacity of the Virgin Islands refinery of Amerada Hess on 12/1/74 as reported by the Federal Energy Administration. The 1951 data are for the present firm or its lineal predecessor. A zero indicates that the company was not a refiner on 1/1/51 and did not become a refiner by acquiring a company that was refining on 1/1/51, although subsequent to its entry into refining it may have acquired such a firm.

EXHIBIT E

VERTICAL INTEGRATION FROM SELECTED INDUSTRIES WITH HIGH AND LOW CONCENTRATION

Locomotives & Parts (4-firm 97%, 8-firm 99%)

Bearings, magnet wire, foundry, electronics and controls, lamps, locomotive parts, financing.

Primary Aluminum (4-firm 96%, 8-firm 100%)

Distribution (mill products), cutlery, building products, bauxite mines, financing, fuels (coal), steamships (for alumina & bauxite), railroads, electrical controls for industrial machinery, hydroelectric power plants, metal working, machinery, magnet wire, electrical products, cooking untensils, packaging and foil, chemicals, truck bodies and trailers.

Flat Glass (4-firm 94%, 8-firm 98%)

Direct sales to customers, railroad electronic systems and other equipment, plastics, chemicals, hoses and fittings for industrial uses, manufacture.

Motor Vehicles (4-firm 91%, 8-firm 97%)

Manufacture, repair parts, retail dealership, financing, transportation (trucks, locomotives, diesel engines), power generating plants, construction equipment.

Chewing Gum (4-firm 85%, 8-firm 97%)

Manufacture, chemicals, import company.

Cigarettes (4-firm 81%, 8-firm 100%)

Manufacture, health insurance, financial services and asset management.

Copper (4-firm 75%, 8-firm 98%)

Mining, metal processing, manufacture of metal products, wire mill products, containers, uranium, timberlands, electronics, building industry products, railroads, steamships, dock and barge companies, coal companies.

Tires and Innertubes (4-firm 72%, 8-firm 98%)

Manufacture, retail stores, finance, rubber plantations, marine transport, insurance.

Metal Cans (4-firm 72%, 8-firm 83%)

Manufacture of cans, wrappers, timberland, chemicals, recovery of tin, production of inorganic tin chemicals, electroplating, organic coatings, inks, machinery and equipment for can manufacturing, waste processing and reclamation and recycling, corregated shipping containers, pulp and paperboard and paper production, installation and operation of can assembly facilities dedicated to customers' needs with financing and supply commitment, ultraviolet curing of printing on metal.

EXHIBIT E (Cont'd

Soap and Detergents (4-firm 70%, 8-firm 79%)

Production, vegetable oil mills, chemical pulp plants, timberland, paper and paper pulp and groundwood mills, research, storage and distribution, animal feed ingredients.

Radio and Television (4-firm 48%, 8-firm 67%)

Radio and television receivers, television picture tubes, semiconductor devices, electronic equipment for radio & TV broadcasting, phonograph records, telefilms, marketing of receiving tubes, antennas, electronic instruments, batteries, service organization (installation and maintenance of equipment), communications & telephone systems, TV dinners, broadcasting networks, television broadcasting stations, AM radio stations, FM radio stations, intercontinental television transmission, book publishing, rents and sells construction equipment, financing of office and factory production equipment, all types of satellites, space power and propulsion, research and development.

Steel (4-firm 47%, 8-firm 65%)

Ore and coal mines, limestone quarries, steel plants, ore ships, steel mills producing drilling equipment, steel drums and pails, residential siding, fabricating and engineering of structures and ships, chemicals and fertilizers, railroads, real estate development.

Shoes (4-firm 28%, 8-firm 36%)

Manufacturing, retailing outlets, adhesives, tanneries, import companies.

Pharmaceutical Preparations (4-firm 26%, 8-firm 43%)

Plants, research laboratories, experimental farms, glass containers, retail outlets, chemicals and plastics.

Newspapers (4-firm 16%, 8-firm 14%)

Publishing, paper products, timberland, ships, distribution, newsprint.

Bottled and Canned Soft Drinks (4-firm 13%, 8-firm 20%)

Manufactures bottles, distributes and sells syrups, concentrates, beverages, transportation, leasing, snack foods.

Sources: Department of Commerce, Census of Manufacturers, 1970 and Moody's Industrial Manual, 1975.

8

Competition in Refining

TESTIMONY BY DONALD C. O'HARA, PRESIDENT, NATIONAL PETROLEUM REFINERS ASSOCIATION, BEFORE THE SUBCOMMITTEE ON ANTITRUST AND MONOPOLY, COMMITTEE ON THE JUDICIARY, UNITED STATES SENATE — NOVEMBER 12, 1975.

Summary: The oil refining industry is not concentrated; the largest single refiner has less than 10 percent of the total refining capacity; the share held by the majors has declined; there has been entry into the "major" class and there are a large number of strong independent competitors. The monopoly of the old Standard Oil group was eroded more by the discoveries of new oil fields and the advent of the automobile than by the 1911 decree. Mostly the integrated units flourished after the divestiture. There will be administrative problems created by the government's attempting to define "refining." We need new energy sources and most of the R&D in these fields is done by the largest refiners. Divestiture would hinder these efforts and make us more dependent upon foreign imports.

Donald C. O'Hara speaks . . .

My name is Donald C. O'Hara. I am president of the National Petroleum Refiners Association, which has as its members substantially all of the companied engaged in refining or petrochemical manufacture in the United States.

In the Congressional debate on the proposed legislation to prohibit oil companies from operating in all phases of the oil business, the principal reason given is that it would make the industry more competitive. Therefore, I should like to address myself to two questions: One, is the petroleum refining industry competitive? And, two, would divestiture make it more competitive?

I. Is the petroleum refining industry competitive?

One subject that appears frequently in the Congressional debate is reference to the top twenty companies as "majors." For example, in his statement before your Subcommittee on October 29, Congressman James V. Stanton of Ohio said, ". . .it is generally recognized that the industry is dominated by twenty 'majors.' " What is not generally recognized is that this group of twenty is not a closed club. Several aggressive competitors have entered this group in recent years, elbowing out established companies. I have here a chart marked Exhibit 1 which shows the principal U.S. refining companies in the order of their capacity in 1951. I have used 1951 because in 1951 the Department of the Interior published a chart listing all of the U.S. refining companies in the order of their capacity and classifying the top twenty (those with 50,000 barrels per day or more of refining capacity) as "majors." This is the list which appears in the left hand column of Exhibit 1. The right hand column lists the refining companies with a capacity of 50,000 barrels per day or more on January 1, 1975.

EXHIBIT 1

U.S. REFINING COMPANIES
WITH 50,000 BARRELS PER DAY OR MORE CAPACITY
(numbered in the order of their capacity)

January 1, 1951		January 1, 1975	
Total U.S. capacity	6,963,644	Total U.S. capacity	15,770,050
1. Standard (N.J.)	769,500	1. Exxon	1,243,000
2. Socony Vacuum	548,500	7. Mobil	903,500
3. The Texas Co.	510,000	4. Texaco	1,073,000
4. Standard (Indiana)	502,500	3. Standard (Indiana)	1,115,000
5. Gulf	458,500	6. Gulf	904,200
6. Standard (California)	388,200	5. Standard (California)	952,000
7. Shell	382,000	2. Shell	1,150,000
8. Sinclair	360,000		
9. Cities Service	215,000	17. Cities Service	268,000
10. Sun	200,000	10. Sun	569,000
11. Tidewater	192,500	18. Getty	218,731
12. Phillips	182,500	13. Phillips	408,000
13. Atlantic	167,000	9. Atlantic Richfield	724,600
14. Union	153,400	11. Union	496,000
15. Standard (Ohio)	116,500	12. Standard (Ohio)	431,000
16. Pure	113,300		
17. Richfield	110,000		
18. Ashland	105,700	15. Ashland	362,943

19. Continental	93,150	14. Continental	377,000
20. Texas City	50,000	31. Texas City	74,500
Total	5,618,250	Total	11,270,474
Percent of U.S. total	80.68%	Percent of U.S. total	71.47%
		8. Amerada Hess	797,900
		16. Marathon	324,000
		19. Coastal States	212,982
		20. Petrofina	200,000
		21. Kerr-McGee	166,000
		22. Commonwealth	161,000
		23. Union Pacific	152,200
		24. Murphy	137,500
		25. Koch Industries	109,800
		26. Clark	108,000
		27. Tenneco	103,000
		28. Crown Central	100,000
		29. Toscopetro	87,000
		30. Charter	85,900
		32. Farmland	73,838
		33. Tesoro	64,000
		34. Pennzoil	62,600
		35. Hawaiian Independent	60,000
		36. Husky	59,000
		37. Apco	58,670
		38. United Refining	58,000
		39. Nat'l. Coop. (NCRA)	54,150

There have been several significant changes in the intervening years. The refining capacity of the 1951 "majors" has doubled from 5.6 million barrels per day to 11.2 million barrels per day, but their percentage of total U.S. capacity has *declined* from 80.68% to 71.47%. In fact, it has dropped five points in the past two years. There have been four newcomers to the top twenty—Amerada Hess, Marathon, Coastal States and Petrofina—and eighteen additional companies with 50,000 barrels per day or more of capacity.

To put this in perspective, the automobile industry talks about the "big three." Other industries talk about the "big four" or the "big five." This chart shows that in the oil refining industry: (1) the largest single refiner has less than ten percent of the total refining capacity; (2) the share held by the major companies has declined; and (3) twenty-two newcomers have built enough new capacity to put them in a class with the 1951 majors.

The fact is that no other basic industry even comes close to having the large number of strong independent competitors that we have in the oil refining business.

There seems to be a great deal of misunderstanding about this point. For example, in the debate in the Senate on October 7, 1975, Senator Hart of Colorado quoted from a study by Professors Fred C. Allvine and James M. Patterson, of the University of Indiana, in which they said:

> "Since 1950 the integrated oil companies have taken over several of the important independent refineries and there have been built no new independent refineries with over 50,000 barrel per day capacity." (121 Cong. Rec. 17690, daily ed. Oct. 7, 1975)

In fact, the following seven new refineries with a capacity of 50,000 barrels per day or more have been built by independent companies since 1950:

Company		**b/d capacity**
Amerada Hess	St. Croix, V.I.	700,000
Commonwealth	Puerto Rico	161,000
Koch Industries	St. Paul, Minn.	109,800
Amerada Hess	Perth Amboy, N.J.	67,900
Hawaiian Independent	Honolulu, Hawaii	60,000
United Refining	Warren, Pa.	58,000
ECOL	New Orleans, La.	200,000 (under construction)

In addition, fifteen other independent companies have each built 50,000 barrels per day or more of new capacity since 1950, either by adding to existing refineries or by building on a site previously occupied by an older refinery. This means that actually the equivalent of twenty-two additional refineries of 50,000 barrels per day or more have been built by independent companies since 1950.[1]

Even this does not tell the whole story. At least sixteen independent companies have tried unsuccessfully to get building permits to build refineries on the East Coast. These applications have ranged all the way from the Canadian border in Maine to the Virgin Islands. Some of them, such as the Steuart Petroleum Company here in Washington and the company owned by the late Aristotle Onassis in New Hampshire, have been given wide publicity. Others are well known in the industry. The chief obstacle to the

[1]The latter group also is examined in Exhibit B of W. R. Peirson's testimony, page 109.

entry of new independent refiners in the refining business has been the objection of the local communities.

And there is still more to the complete story. As a result of the fact that companies cannot get permission to build on the East Coast, some independent companies have built refineries immediately offshore to serve the American market. These refineries have been built in the Caribbean Islands and in Newfoundland and Nova Scotia. These constitute a substantial entry of new independent competitors in refining.

II. Would divestiture make refining more competitive?

Another favorite subject of the Congressional debate is comparing divestiture with the 1911 decree of the Supreme Court which broke up the old Standard Oil Company.[2] We are in a unique position to comment on this because our Association was founded in 1902. At that time there were only two factions in the oil business—the Standard Oil group and those who were outside of it. Our Association was formed by the outsiders. When I came to work for the Association at the end of World War II, we still had on our Board of Directors a half dozen men who had lived through that period. Some of them were witnesses in the case against Standard Oil.

There are several important factors in that case that most people have forgotten. The first is that the 1911 decree did not have any important impact on competition until many years later. The records of our Association show that in the early 1920's—more than ten years after the decree—we were involved in the litigation before the Interstate Commerce Commission over the control of tank cars because the companies in the Standard Oil group continued to work together until the stock gradually passed into other hands.

You can picture what would happen today if you divorced a pipeline running from a producing field to the gate of a refinery designed to process that crude. What do you think the buyer would do with the line? The answer is obvious. The line could not be sold or operated unless the new owner got a long term contract from the refiner. Thus, regardless of ownership, the line would continue to be a captive of the refinery. What the law would do is prevent the refiner's competitors from building a line.

Secondly, the monopoly of the Standard Oil group was eroded by two important developments in the 1920's—the discovery of new oil fields, principally in Texas, Oklahoma and California, and the rapid development of the automobile with its appetite for gasoline which created a huge new

[2]For more background on the Standard Oil breakup of 1911 see Chapter 5 by Hastings Wyman.

market. These changes opened up the oil industry to the entrance of new competitors. Real competition in refining resulted from these economic changes and the enforcement of the Interstate Commerce Act more than it did from the 1911 decree.

Thirdly, there is a widespread belief that all of the companies flourished after the divestiture. The fact is that those that were integrated units flourished, but others did not. For example, one of those cut off was the Vacuum Oil Company, one of the leading refiners of the day. It found that it could not survive without marketing outlets and in 1931, twenty years after the decree, it merged with the Standard Oil Company of New York, one of the other companies divorced by the 1911 decree, to form the Socony-Vacuum Oil Company (now Mobil). This was challenged in a court action as a violation of the decree, but the court approved it as a matter of economic necessity.

Lastly, and this is particularly important in the present case, the 1911 decree separated companies which were complete legal units, even though they were not all complete economic units. The various subsidiaries of Standard Oil were freed from common control of their stock, but there was no legal problem of identifying the assets of each company. For a time they were restricted to their own marketing territory by the fact that they could not use the brand name "Standard" outside of that territory but they invaded each other's markets with other brand names. The Standard Oil Company of California, for example, competes nationwide with its brand "Chevron." The New York company uses the name "Mobil," etc.

The pending bills would prohibit the divorced companies in production or marketing, for example, from entering the refining business. This would have the opposite effect of the 1911 decree because it would keep the most likely competitors out of the business.

The pending bills propose to prohibit the companies from engaging in a particular activity, i.e., "refining" but no one knows how to define that activity. It may surprise you to know that government agencies have never been able to agree on a definition. Refining is the process of manufacturing products from hydrocarbon feedstocks. These include not only well known products such as gasoline, jet fuel, heating oil and residual oil, but also many others such as lubricating oil, petrochemicals, etc. About two-thirds of all textiles and all automobile tires are made from petroleum. These products are usually made from crude oil, but they are also made from natural gas liquids, shale oil, coal, tar sands and gilsonite. They often are made from petroleum fractions such as naphtha which are imported into the United States after the heavier fractions such as residual oil have been removed in a foreign country.

Because of the complex nature of refining and its interrelationship with petrochemical manufacture, the proposed legislation would require years of litigation before its limits could be defined.

(1) Three refineries are not on the Bureau of Mines' list because they are outside the customs territory of the U.S.—in the Virgin Islands, in a foreign trade zone and on the Island of Guam. The Oil Import Administration, now a part of the Federal Energy Administration, says that these are not U.S. refineries and it does not grant them allocations to import oil. Another division of the FEA says that they are U.S. refineries and allocates domestic oil to them under the entitlements program. In fact, the FEA has granted entitlements to another U.S. company which is the principal owner of a foreign refinery.

(2) Several other plants do not appear on the list because the Bureau of Mines says that they are not refineries. As a raw material they use naphtha, which has already been partly refined in a foreign refinery. Since the heavy part of the barrel has been removed, these refineries have a gasoline manufacturing capacity which is about twice that of a refinery processing crude oil. If crude oil is shipped from the U.S. to a foreign refinery on a Caribbean island (as was done in at least one recent case),this would not be covered by the proposed bills. Furthermore, if the crude oil were separated into residual oil and naphtha and returned separately to the U.S., the naphtha could be made into gasoline or petrochemicals in the U.S. without meeting any government definition of "refining."

(3) The Bureau of Mines' refinery list does not include natural gas plants, many of which make a product that can be marketed directly as jet fuel or that can be blended in a refinery or petrochemical plant without going through the crude unit and, thus, are not counted as refinery inputs.

(4) There are forty petrochemical plants which receive allocations to import foreign oil in the same way that refineries do, i.e., based on their inputs of feedstocks which are mostly natural gas liquids or naphtha. The Bureau of Mines says that they are not refineries.

(5) Some time ago a chemical company in Michigan purchased an oil refinery. Although the refinery continued to supply gasoline to its retail dealers, the Oil Import Administration classified the plant as a chemical plant and denied it an import allocation. This decision was later reversed.

(6) There are sixteen refineries on the Bureau of Mines' list which did not receive an allocation to import oil.

(7) There are five plants which received an allocation to import oil which do not meet the Bureau of Mines' definition of a refinery.

In summary, we do not believe that it is possible to define "refining" or that it is necessary to define it. At the present time, oil companies are free to make petrochemicals, natural gas plants are free to make jet fuel, etc. No one is insulated from competition. As our Exhibit A shows, new competitors are constantly entering into the refining business.

The proposed legislation would prohibit oil companies from "refining." This would plunge the whole manufacturing and distribution system of the industry into chaos while the courts and government agencies decide what is "refining." The history of government controls in recent years is replete with cases in which refineries are classified as foreign by one government agency and domestic by another; as refineries by one department and not as refineries by another (or another branch of the same agency). This would accentuate a process which began under the Oil Import Program and advanced with the FEA—success in the refining business will no longer come to those who make a product the customers want; it will come to those who have the political influence to get a favorable ruling from the government.

Conclusion

Divestiture has been proposed by some as a benefit to the independent refiners. I will repeat what I told this Subcommittee in August, 1974: "I am personally acquainted with nearly every independent refiner in the U.S., and I do not know a single one who supports the theory that any benefit would accrue to the independent refiners by breaking up the large companies." Nothing that has happened since that time has changed this situation.

This does not mean that independent refiners do not have some quarrels with the larger companies. I have read the statements that have been made by some independent refiners before your Subcommittee during the present hearings. At the risk of oversimplifying a complex subject I would summarize their views as favoring: (1) that integrated companies operate each function of their company profitably; and (2) that there should be a competitive market in crude oil.

We have no hesitation in endorsing these views in principle. In the first case, several major companies have already moved toward setting up separate corporations for each function. In the second, we favor a competitive market in crude oil but we believe that the proposal made by one group to make sales through a government corporation is unnecessarily complicated and would involve extensive government regulation.

The underlying cause of our present energy crisis is the fact that our domestic production of oil is declining while our consumption of oil is increasing. This makes us increasingly dependent upon foreign imports. We have heard a great deal about plans to make our country more independent. We believe that divestiture of refining would have the effect of making us more dependent upon foreign imports.

One of the principal hopes for energy independence is the possibility of alternate sources of energy, of which shale oil, coal and tar sands are the most promising. Most of the research and development work in these fields is being done by the companies which are our largest refiners. They have the expertise and the facilities to carry forward these experiments. A law forbidding these companies to engage in the manufacturing business would be a setback to our hopes for energy independence.

A second major objective in our drive for independence has been to increase our domestic refining capacity. At the present time we import between 10 and 15 percent of our total consumption of petroleum products (in addition to crude oil imports). The situation is even more marked on the East Coast, which imports about 25 percent. A few years ago we were told that it was government policy to encourage the building of domestic refineries and, on April 18, 1973, the President issued a Proclamation at the request of the chairman of the Oil Policy Committee establishing a graduated scale of supplemental fees on finished products. These fees were to be increased over a period of years. The purpose, according to the Proclamation, was:

> "...in order to discourage the importation in the United States of petroleum and petroleum products in such quantities or under such circumstances as to threaten to impair the national security...to increase the capacity of domestic refineries and petrochemical plants to meet such requirements; and to encourage investment, exploration and development necessary to assure such growth."

Now it appears that we have turned in the opposite direction. The supplemental fee on petroleum products has been lifted and Congress is considering legislation to divorce domestic refineries from their supply of raw material. Such legislation, of course, would not apply to foreign refineries, which would be thus given a substantial competitive advantage over domestic refineries and petrochemical plants.

9

Vertical Divestiture: Pipelines

STATEMENT OF CHARLES J. WAIDELICH, PRESIDENT, CITIES SERVICE COMPANY, BEFORE THE SUBCOMMITTEE ON MONOPOLIES AND COMMERCIAL LAW, HOUSE JUDICIARY COMMITTEE — SEPTEMBER 10, 1975

Summary: Economic and financial considerations dictate joint ventures; no one company produces the very high volumes of crude or product necessary to fully utilize lowest cost, large-diameter pipelines. Under the 1941 Consent Decree, dividends on pipeline investment which oil companies can receive are restricted. This has tended to hold down shipping rates and pipeline earnings. After divestiture, this restraint would no longer exist, with a probable increase in rates resulting; this would increase product prices to the consumer.

Charles J. Waidelich speaks . . .

I believe it advisable first to distinguish petroleum product pipelines from crude oil pipelines. Whereas crude oil pipelines transport crude oil from producing fields and ports of import to the refining areas to be refined, petroleum product pipelines transport the refined products from the refineries and ports of import to the consuming markets. Generally speaking, petroleum product pipelines are long-distance, large-volume common carriers of refined products. The refined products carried include things such as gasolines, heating oils, diesel and jet fuels.

Experience has demonstrated that pipelines are the most efficient and inexpensive overland means of transporting petroleum products in addition to being environmentally superior to other modes of transportation. As a result, product pipeline transportation costs represent a small but important component of delivered petroleum prices, perhaps in the range of 2 to 3 percent.

Historically, oil companies, faced with the need to move petroleum considerable distances at the lowest cost possible, have undertaken to build and operate this nation's complex system of pipelines. As America's needs for petroleum products have soared, the industry has continued to expand this pipeline system as the demands of the marketplace dictated.

Today there are over 200 liquid pipeline companies in the United States, a multi-billion dollar industry operating over 200,000 miles of crude oil and petroleum product lines. A large majority of this pipeline mileage is operated by petroleum companies, large and small.

Today more than ever before, continued expansion of the pipeline industry is a priority item. Private industry has met the challenge in the past. Permitted to do so, it will continue to fulfill this country's continually expanding need for efficient transportation of petroleum products at lower costs which must benefit the consumer.

Most Petroleum Product Pipelines are Joint Ventures

Most petroleum product pipelines are joint ventures because the capital requirements are so large that very few, if any, corporate entities are able to dedicate the substantial amount of cash or debt obligations required to construct a pipeline. Joint ownership of pipelines by oil companies is a practice developed from necessity which has existed in the industry for many years. Substantially all long, large-diameter pipelines have been conceived, constructed, and owned by groups of oil companies. Economic and financial considerations dictate this form of ownership. Economic necessity has required that many petroleum pipelines, if they are to compete with tankers, trucks and other modes of transportation, be large-diameter lines.

Pipelines have a high original investment and a relatively low annual operating cost. Construction costs increase at a much slower rate than the potential carrying capacity of the line. For example, the cost of constructing a 36-inch line is less than 3-1/2 times the cost of a 12-inch line, yet the 36-inch line can carry 17 times the volume of the 12-inch line. Phrased another way, one would have to construct 17 pipelines, each 12 inches in diameter to carry the same volume as one 36-inch line. The cost of operating a 36-inch line on a per-barrel basis is about a third the cost of operating a 12-inch line.

To illustrate the magnitude of investment required, Explorer Pipeline is a large diameter pipeline (mostly 24 inches to 28 inches) extending from the Gulf Coast to Tulsa, Oklahoma, and then on to the Chicago area. The entire completed pipeline system required an investment of approximately

$219 million, and these were 1970 and 1971 dollars. None of the non-oil-company-owned pipelines approach in magnitude the costs and capacity of the large-scale joint venture lines. Only five of the pipelines owned by non-oil interests exceed 500 miles in length. Two of these, Buckeye Pipeline Company and Southern Pacific Pipeline Company, are owned by railroads.

Williams Pipe Line Company is the most extensive pipeline system owned by independent interests. In 1965, Williams Brothers acquired the Great Lakes Pipe Line System which began operation in 1931, and was constructed and operated by a group of oil companies until the acquisition by Williams Brothers. Williams Pipe Line Company thus began its independent existence as a pipeline already successfully operated for 34 years with firmly-established, long-term customers. Its acquisition occurred after the obligations of the original owners under their throughput obligations had been satisfied and the pipeline's original debt retired.

Since World War II, there have been a number of attempts to promote big-inch, that is, large-diameter, pipelines as privately-owned speculative ventures. All failed for lack of financing. In 1951, an application was filed by the U.S. Pipe Line Company for priority assistance to obtain steel to construct a 563-mile, 16-inch products pipeline from Beaumont, Texas, to Cincinnati, Ohio, costing $42,600,000. Priority was at first refused, but after the project was amended to extend the pipeline to Newark, New Jersey, at an increased cost of $142,000,000, the steel allocation for constructing the line was granted. The Petroleum Administration for Defense also recommended that a Certificate of Necessity for accelerated amortization be granted at a 25% write-off. Nevertheless, this project failed for lack of financial backing.

In 1953, American Pipeline Corporation (no connection with any oil company) filed an application with the PAD for allocation of steel and a Certificate of Amortization for the construction of a 1,425-mile, 22- to 26-inch products line from Beaumont, Texas, to Newark with a capacity of 500,000 barrels per day and a total cost of $156,000,000. After the project was amended to increase the line size to 30 to 32 inches to raise the capacity to 1 million barrels per day, a Tax Certificate was issued allowing increased tax amortization. This project also failed for lack of firm commitments from prospective shippers. The proposed project was revived during the Suez crisis. At that time, the Office of Defense Mobilization considered but rejected government aid for financial backing although the military need for the line was recognized.

In the early part of 1956, the West Coast Pipeline Company filed an application with the PAD for allocation of steel to construct a 20- to 26-inch, 1,030-mile crude oil line from Midland, Texas, to Norwalk, California.

Steel allotment and rapid tax amortization were both granted, but the line could not be financed because of lack of throughput commitments.

With but few exceptions, petroleum pipelines have been built by oil companies because they alone have had the sources of supply and established markets necessary for the line's successful operations and because they have been the only investors willing to venture the requisite risk capital and willing to undertake the throughput obligations necessary to obtain the required financing. In short, construction of petroleum pipelines by oil companies was essential to the maintenance and further development of a national pipeline system adequate to meet our growing demand for petroleum products.

An appreciation of various facets of the oil and product pipeline industry as it exists today is necessary to understand why it is still desirable, if not absolutely necessary, that the oil companies have a financial interest in pipeline operations. First, and of major importance, the operation of an oil pipeline historically has been proven to be a capital intensive undertaking involving risks. Investment in a pipeline is inflexible and cannot be changed to respond to changing patterns of distribution as can most other types of transportation systems.

Pipelines transport only one kind of material, liquids, over a fixed route and in one direction; if business falls off due to changing market conditions or changing supply locations such as new refinery construction, foreign imports into markets and the like, it is unlikely that a pipeline will be able to sustain profitable operations. Changes in conditions at either end of the line and even along the line may adversely affect operations. Because of the high percentage of fixed costs, the profitability is dependent upon the maintenance of constant throughput at projected levels. Losses can occur with slight fluctuations in the volume of throughput.

In this connection, I refer you to the testimony of Vernon T. Jones, then President and Chief Executive Officer of Explorer Pipeline Company, given before the Subcommittee on the Judiciary on January 29, 1975. There, Mr. Jones stated that during the period October 1971 to year-end 1975, Explorer will have accumulated losses of over $28,000,000, and during that interval Explorer has made calls for cash advances from its stockholders totaling $25,500,000. He stated those advances were made the terms of the Throughput and Deficiency Agreement between the stockholders, which Agreement is the basic security for Explorer's debt. Mr. Jones made those statements in January of this year. I am advised that through June and August of this year, those figures are in fact now in excess of $35,000,000 and $30,000,000 respectively.

Traditionally, the financing of pipelines by their owners has been through the private placement of 30- to 40- year notes held by many lenders. The debt is secured by a throughput agreement of the owners of the pipeline containing cash deficiency obligations which in essence guarantee the pipeline company's indebtedness. It is doubtful that pipeline companies alone could obtain financing at comparable costs because of the lack of an underlying throughput obligation or guarantee by its owners.

It has been suggested that perhaps the large oil companies should commit themselves to throughput obligations for the benefit of independent entrepreneur-owned pipelines. I rather doubt that the stockholders of my Company would be enthused about the idea of lending our credit to an entity in which the Company has no equity interest and no voice in its operations. I understand that there are some companies which have outstanding indentures which specifically prohibit guarantee of the indebtedness of another entity in amounts other than the guarantor's percentage of ownership in that entity.

I do not want to leave you with the impression that our Company, for instance, would not ship product over pipelines in which it would own no interest. We frequently do so over such lines as Buckeye Pipeline, Plantation Pipeline and Williams Pipeline companies. We have no ownership interest in those lines. Further, should non-oil interests propose a new viable and competitive pipeline project and poll Cities Service to determine if we would ship product over such a line, rest assured we would give an affirmative answer if we had a need to move product along the route of the line, and if the tariffs would be competitive with other means of transportation available to us.

Today more than ever it is in the public interest to allow and even encourage the development and operation of oil and product pipelines. Probably no other form of overland transportation has such a proven record of effiicency, safety and environmental soundness.

It is well known that under the 1941 Consent Decree, the dividends on pipeline investment which oil companies can receive is restricted. Consequently, this has tended to hold down shipping rates and pipeline earnings. If divestiture were required, this restraint would no longer exist, and there likely would be an increase in shipping rates with a corresponding increase in the price of petroleum products to the ultimate consumer.

In addition, regulation of pipeline companies as common carriers requires that all shippers, whether or not they are owners of the pipeline, have an equal right to have their products transported and at the same tariff.

As opposed to being anticompetitive, the facts show that joint venture pipelines have provided positive benefits for the independent shippers as well as the consumer.

Let me say a further word about throughput agreements. There appears to be a popular misconception that if a shipper-owner has, for example, 10 percent of the stock of a joint venture products pipeline, the owner is entitled to 10 percent of the available space in the pipeline. This is not true. The experience of Colonial Pipeline can serve as a good example of what actually happens.

Fortunately, Colonial has been operating substantially at full capacity for some time. This includes substantial shipments made by non-owners. The pipeline is a common carrier, and it must accept all properly tendered shipments including those tendered by non-owners. If the line is full, it follows that the percentage of capacity utilized by its shipper-owners must be reduced. They must "move over" to accommodate the shipments of the non-owners. This is the typical pipeline proration situation. When the line is full, each would-be shipper, including the owners of the line, must reduce the quantities it ships on an equitable prorata basis. This situation is likely to be a temporary one because if the demand is there, the line will be expanded.

A typical throughput agreement provides that each shipper-owner will ship or cause to be shipped by the pipeline, its percentage of the amount of petroleum products which, with the amounts (if any) shipped by non-owners will be sufficient to provide the pipeline at its published tariff rates with a gross amount of cash revenue at least sufficient with other available cash resources of the pipeline to pay and discharge all of the pipeline's expenses, liabilities and obligations as they become due. This is the typical "ship or pay" provision.

It is conceivable that Owner "A", with a 10% interest in the pipeline, might ship quantities sufficient to generate 20% of the cash required, while Owner "B", also with a 10% ownership, could have nothing shipped; however, since sufficient cash would have been generated to meet current obligations, a deficiency situation would not exist. In other words, throughput and deficiency in the cash required to meet all expenses, liabilities and obligations; but when there occurs a deficiency, such as has occurred with Explorer Pipeline, payments must be made by the owners.

I cannot emphasize too strongly that the typical throughput agreement is an obligation. It does not entitle the owner, as a shipper, to any special service or consideration. It merely constitutes the guarantee of the shipper-owners of their respective proportions of the debt of the pipeline.

Let's briefly compare some of the risks and obligations of owners of the pipeline with the risks and obligations of non-owners. The owners have effectively guaranteed that the pipeline debt will be repaid whether or not the pipeline is completed, successful, or abandoned. Non-owners have no such obligation. Looking briefly at some of the benefits enjoyed by non-owners, we find they have no investment and are free to use their resources for other purposes, they have no debt guaranteed by them, they enjoy freedom of change, and they may ship with the same rights to access to the pipeline and at the same tariffs as owners.

How Joint Venture Pipelines are Conceived and Planned

This is a complex matter which usually begins in the mind of some demographer in an oil company's long-range planning department who believes that population growth in a particular part of the country will require increased volumes of petroleum products for consumers in some reasonably foreseeable time. Or the idea might be conceived by a marketing department which believes that through aggressive marketing and increasing consumer acceptance it might increase the volumes which it can successfully market in a particular area. Analyses at this stage of planning would almost inevitably indicate that no one company would have sufficient new volumes or volumes it could switch from other areas to the area in question to justify the cost of a modern, efficient pipeline with an expandable capacity to meet future needs.

At this stage then, a study is made of whether other companies might be interested in increasing their product availability or efficiency in moving existing volumes in the area concerned. These other companies are then approached, usually on a very informal basis, to see if they have an interest in the project. There would be a number of conversations, and if any real interest were expressed, there would be a meeting of what would probably be called a Management Committee of prospective shippers-owners. This Committee would initially appoint a Feasibility Study Committee.

At this stage of the planning, the group would include those companies which from the public record would appear to have an obvious use for the proposed facility. If after appropriate study the new facility is deemed feasible, there would be appointed a Right-of-Way Committee to plan the proposed facility on paper, and an Engineering Committee to do preliminary engineering studies, including studies of alternate line sizes, considering the possibilities that others would join as shipper-owners in the project and allowing for the estimated usage by non-owner shippers. If the project then appears feasible, there will be appointed a Finance Committee and a Legal Committee, the former to ascertain on an informal basis if funds will be available to construct the line and upon what basis.

History indicates that financing will probably be available if the lenders are assured that the financial strength of the owners is behind the project by means of the customary throughput and deficiency agreement. Also, additional solicitation is made of those persons that the group thinks would be logical shipper-owners. There follows, then, the customary pattern of documentation, including a shareholders' agreement and the all-important throughput agreement.

Perhaps a word should be said about the shareholders' agreement. This document provides, generally speaking, for representation on the board of directors by members of the group in proportion to their ownership interest which is generally based upon the percentage of throughput they are willing to guarantee. It frequently provides also for a right of first refusal in the remaining members of the group to purchase the stock which any member of the group may wish to sell.

It is fairly fundamental that no selling throughput obligor would want to sell its interest in the facility to a less than responsible third party when it would still be bound by its obligation under the throughput agreement. Thus, it is important not only to the selling shareholder, but also the the remaining shipper-owners that they be satisfied that a substitute owner has the ability, resourcefulness and willingness to assume the throughput guarantee obligations of the selling shareholder. Of course, the lenders might well have a word or two to say about discharging one shipper-owner from a throughput obligation and substituting another.

Joint Ownership by Major Companies

Why do the larger oil companies build their product pipelines as joint ventures rather than as wholly owned entitites? First, it is highly unlikely that any oil company would have the very high volumes of product to ship in order to achieve the lowest cost and the most efficient transportation resulting from large-diameter pipelines. It is not the joint venture itself which the companies seek, but the lowest cost and most efficient transportation. This means large-diameter, high-volume pipelines.

Only through joint ventures may sufficient volumes of product be accumulated to justify large-diameter pipelines. Only those companies needing to move product from specific supply locations to specific marketing locations are interested in such pipeline transportation. Their desire to find the lowest cost transportation ultimately benefits the consumer. Second, there is the question of risk. No prudent company would willingly risk the large amounts of capital required in a single project of this type. Capital is in short supply and that which is available must be prudently diversified.

10

Pipelines, Divestiture and Independents

STATEMENT OF FRED F. STEINGRABER, PRESIDENT AND CHIEF EXECUTIVE OFFICER, COLONIAL PIPELINES, BEFORE THE SUBCOMMITTEE ON ANTITRUST AND MONOPOLY, COMMITTEE ON THE JUDICIARY, UNITED STATES SENATE — JANUARY 30, 1975.

Summary: Independent producers, who have increased their market share in the last decade, have never supported divorcement of pipelines from oil companies. The pipelines are adequately regulated by present laws under the ICC, with the entire system available to all potential shippers, large or small, owners or nonowners. Divestiture would increase financing costs since any new project would have to compete in the market for speculative money, and could not take advantage of the low interest rates and long-term private financing made possible by parent company financial commitments. Rates would have to rise. Some service would be abandoned if it is near depleted fields where the carrier finds service no longer economical.

Fred F. Steingraber speaks . . .

During the first series of hearings held before the Subcommittee on Antitrust and Monopoly of the Senate Judiciary Committee, August 6-8, 1974, on the Industrial Reorganization Act (S. 1167) there were several questions raised about oil pipelines which I feel require and deserve further comment.

Senator Hart, in his opening statement, referred to the control of pipelines by major oil companies and its adverse effect on independent oil companies and on the nation. As a matter of fact, refiners and marketers of petroleum products, as well as consumers, have materially benefited from the network of oil pipelines. More sellers of petroleum products are able to reach more different markets more efficiently and at a lesser cost and with greater quantities of products by these pipelines than would otherwise be possible. Pipeline critics often repeat the myth that the ownership of pipelines by oil companies harms the independent producer or marketer.

According to the Office of the Energy Advisor,[1] independents, instead of losing ground, have steadily increased their market percentages in the last decade, both in the refining and marketing areas. Their studies reveal that the percentage of independents engaged in refining increased from 16.1 percent to 17.1 percent in Districts I-IV and from 11.1 percent to 13.6 percent in District V during the ten year period from 1960 to 1970, and that the independents' share of the gasoline market increased from 19.81 percent to 25.44 percent in the five-year period from 1968 to 1972.

There are 100 pipelines subject to ICC regulation which substantially make up the pipeline network in the United States. Studies made by the Association of Oil Pipe Lines from published data of the Interstate Commerce Commission for the year 1973 reveal that of these 100 companies: 38 are owned exclusively by the major oil companies; 46 are owned exclusively by independents; and 16 are owned by joint venture groups that include participation by both majors and independents. I might add that included in the "independent" grouping are wholly-owned pipeline subsidiaries of such large and substantially solid firms as the Atchison, Topeka and Santa Fe Railroad, the Southern Pacific Railroad and the Texas Eastern Transmission Corporation.

Viewed in terms of mileage, the 38 pipelines owned by the major oil companies comprise 66 percent of the total mileage of the pipeline system; the 46 pipelines owned by the independents comprise 26 percent of that system; and the 16 owned by joint venture groups of majors and independents comprise eight percent of the system.

The oil pipeline industry is doing and has done an outstanding job for all of the producers and refiners who wish to ship over the pipeline system, whether they are small or large shippers and whether they are owners or non-owners of the pipeline used. I am completely convinced that:

1. It would be the height of *unwisdom* for this Subcommittee to recommend legislation designed to compel oil companies to divorce themselves of ownership of oil pipelines.

---

[1]Staff Analysis, Office of the Energy Advisor, Department of the Treasury, dated August 27, 1973.

2. Oil pipelines have performed and are performing a valuable transportation service at a uniquely low cost to the so-called independent segment of the petroleum industry.

3. Present regulation of the oil pipelines by the ICC is soundly conceived and administered and has been an important factor in the development of this country's pipeline system.

The integrated ownership of pipelines affords a number of advantages to small crude producers which they recognize and appreciate. These benefits include the rapid building and extension of pipelines to areas of newly discovered production, the establishment and maintenance of rates for gathering and trunkline service from areas of low production on a basis of equality with areas of flush production, and the continuation of pipeline service to stripper wells long after transportation has ceased to be profitable. Independent producers have never supported divorcement of pipelines from oil companies.

A letter dated July 12, 1973, addressed to 100 Senators by Mr. Tom B. Medders, Jr., President of the Independent Petroleum Association of America, an association representing 4,000 independent producers of oil and gas in every producing region of the United States, stated:

"It is our understanding that the Senate will vote on Tuesday, July 17, on a modified version of Senator Haskell's amendment to the Alaska Pipeline Bill to require divestiture by 1981 of any ownership or interest in common carrier pipelines by companies which produce crude oil.

"This proposal, I am told, is based primarily on allegations in a Federal Trade Commission staff report which erroneously charges that the independent crude oil producers may have great difficulty securing shipment of their oil, and are subject to discrimination by pipeline companies.

"This is to advise you that we are not aware of any producer having difficulty selling or moving his crude oil, and we do not believe any such discrimination exists.

"Adoption of this amendment, therefore, would be of no benefit to independent producers and could be very harmful to some independents who own or have interests in small pipeline systems. Divestiture, we believe, could serve to increase transportation costs for all independents and raise prices for consumers.

"We will appreciate your taking these views into account in your consideration of this proposal."

Professor Edward J. Mitchell of the University of Michigan, in his statement dated August 6, 1974, filed with this Committee on August 7, has related to you the results of tests which he has conducted to measure the degree of monopolization or competition in the Colonial Pipeline market area. This, in my opinion, would represent the effect such pipelines might have on any market areas. The results of these tests are a matter of record with this Committee. Professor Mitchell concludes that, based on his studies to date, "The pattern of gasoline prices and gasoline market shares are consistent with the working of the normal process of competition and inconsistent with the existence of a cartel of Colonial Pipeline owners or of 'major oil companies'."

In light of these facts, it is not at all surprising that the independent oil companies are satisfied with the service they receive from the pipeline network and are strongly opposed to legislation which would require divorcement of the system from its present owners.

The transportation service offered to petroleum shippers in the United States by the oil pipeline industry is the greatest transportation "bargain" existing in our country. This surely has been a factor in the growth of the independent segment of the industry.

Pipelines Are Adequately Regulated by Present Laws

Interstate oil pipelines were made common carriers subject to regulation under Part 1 of the Interstate Commerce Act by the Hepburn Amendment of 1906. The constitutionality of the Hepburn Amendment, which in effect opened interstate pipelines to public use, was upheld by the Supreme Court in 1914.[2]

Under the Interstate Commerce Act, a common carrier pipeline's tariffs must be just and reasonable and equally applicable to owners and nonowners,[3] must be filed with the ICC before transportation of products begins,[4] with written notice to all known shippers, and must be applicable to all shippers on a nondiscriminatory basis.[5] A pipeline company must furnish transportation upon reasonable request to any shipper, must establish reasonable through or joint rates with all connecting common carrier pipelines, and must provide for a just and equitable division of these joint rates with such connecting lines.[6] Reasonable facilities for delivery of

---

[2]The Pipe Line Cases, 234 U.S. 548(1914); The Uncle Sam Case, 234 U.S. 562(1914); Valvoline Oil Co. vs. U.S. 308 U.S. 141 (1939); Champlin Refining Co. vs. U.S. 329 U.S. 29 (1946); Champlin Refining Co. vs. U.S. 341 U.S. 290 (1951).

[3]49 U.S.C. Sec. 1 (4), (5).

[4]49 U.S.C. Sec. 6 (1), (8).

[5]49 U.S.C. Sec. 8(1).

[6]49 U.S.C. Sec. 1(4).

products to connecting lines must be provided on a nondiscriminatory basis,[7] and the concurrence of all lines participating in joint tariffs must be on file with the Commission.[8] The pipelines' operating criteria and regulations and practices for transporting products must be just and reasonable and uniformly applicable to all shippers.[9] A line is expressly prohibited from giving any unreasonable preference or discriminating in any way in the furnishing of services to different shippers.[10]

Violation of the Interstate Commerce Act can be the subject of a proceeding before the ICC, as well as a Federal court action, and is punishable by penalties which include both fines and imprisonment.[11] The ICC has broad remedial powers under the Act should the pipeline fail to conduct its operations as a true common carrier, as required by statute and by the regulations of the Commission. For example, the Commission is empowered to conduct investigations and hearings upon the complaint of a third party[12] or upon its own initiative,[13] and to issue orders setting single and joint rates,[14] dividing joint rates among connecting carriers,[15] suspending newly filed rates for up to seven months pending investigation of their lawfulness,[16] and ordering reparations for damages sustained by shippers due to violations of the Act.[17]

Pipelines were exempted from the necessity of obtaining certificates of convenience and necessity prior to the construction, enlargement, or abandonment of facilities.[18] This has permitted freedom of competition, sadly lacking in the railroad industry.

Such pervasive regulation by the ICC should and does ensure that pipelines will be operated as common carriers in every sense of the word, and that their operations and practices will be fair and reasonable and will place no one at a competitive disadvantage, even though this ensurement is already dictated and made necessary by the forces of competition with tankers, barges, rail tank cars, trucks and other pipelines.

---

[7] 49 U.S.C. Sec. 3(4).
[8] 49 U.S.C. Sec. 6(4).
[9] 49 U.S.C. Sec. 1(4)-(6).
[10] 49 U.S.C. Sec. 2, 3 (1), (41-88).
[11] 49 U.S.C. Sec. 6(10), (8-10), (16) (41-43).
[12] 49 U.S.C. Sec. 13(1).
[13] 39 U.S.C. Sec. 13(2).
[14] 49 U.S.C. Sec. 15 (1), (3), (6), and (7).
[15] 49 U.S.C. Sec. 13(1), 16(1).
[16] 49 U.S.C. Sec. 15(7).
[17] 49 U.S.C. Sec. 13(1), 16(1).
[18] 49 U.S.C. Sec. 1(18), *et seq.*

Since 1969, approximately 15 cases protesting the tariffs of one or more pipelines have been filed and considered by the ICC's Suspension Board. These protests have raised and put in issue questions such as whether the line's valuation base, upon which the rates are based, is excessive and thus permits the carrier to earn an unacceptably high rate of return; whether the line's rates or revenues per mile, when compared to those of other pipelines and other carriers, are excessive; whether new service, storage, loading and unloading charges are excessive or discriminatory; and whether the rates and charges are discriminatory because they grant an undue preference to other shippers, served by the same carrier, in different areas. Two of these protests have resulted in full suspension of the tariff complained of, and two have resulted in partial suspensions. Some of these cases have put in issue the question of access to pipeline facilities by nonowners. Others have involved the competitive interaction between pipeline tariffs and bulk tank car rates. These cases prove that shippers are willing to challenge pipeline rates and practices when they feel it is in their interests to do so and that the ICC acts promptly and vigorously upon such challenges.

Tariffs generally provide that when tendered shipments exceed the capacity of the line, any transportation furnished by the line will be ratably apportioned among all shippers who are then tendering product.[19] I am attaching hereto Exhibit I which details Colonial Pipeline's proration policy which has been applied uniformly to all shippers since July, 1967, when the system first went on proration.

To further protect shippers against discrimination, a pipeline is prohibited from disclosing to others data concerning volumes shipped and their destinations.[20]

The House Subcommittee on Small Business Problems issued a report on October 18, 1972,[21] which called upon the Interstate Commerce Commission to investigate possible violations of Section 7 of the Clayton Act and the extent to which oil pipelines operate as true common carriers, and possibly discriminate in rates and practices.

Chairman George M. Stafford replied to this Committee on February 1, 1973, stating that the ICC had no complaints of such violations and had adequate authority to investigate and punish offenders if such violations should occur. Mr. Stafford's letter read in part as follows:

"As I am sure you are aware, various provisions of the Interstate Commerce Act require all jointly owned interstate oil pipelines shipping products

[19]49 U.S.C. Sec. 3 (1).

[20]49 U.S.C. Sec. 15 (11).

[21]H.R. Rep. 1617, 92nd Congress, 2nd Session (1972).

of the owners to operate as common carriers without discrimination against any shipper, whether a part owner or an independent, whose products they must also transport. (See e.g., Sections 1(1)(b), 1(3)(a), 1(4-6), 2, 3, 4(1), 5(1), 5a(2), 6, 15(1, 3), and 15a of the Interstate Commerce Act.) Any pipeline that violates these provisions of the law is subject to civil liability for damages (Section 8), and any injured person may complain to this Commission or bring a legal action to recover damages resulting from such violations in the Federal District Courts of the United States (Sections 9 and 13). Certain wilful violations are subject to fines up to $20,000 and two years' imprisonment for each offense (Sections 10, 15(12), and 41(1)).

"Under normal procedure, the Commission institutes investigations upon receipt of complaints that the rates and practices of pipelines are unreasonable and discriminatory under provisions of the Interstate Commerce Act. Under existing statutes, this Commission has adequate authority to compel the attendance and testimony of witnesses (Sections 12 and 18 of the Interstate Commerce Act and Section 41(3) of the Expediting Act (40 U.S.C. Sec. 41(3)), and to take such corrective action as may be necessary (Sections 14, 15, 15a, 16 and 23).

"Historically, our files reveal very few complaints with respect to alleged discriminatory rates and practices of pipelines. Of course, if factual circumstances warrant, an investigation could be instituted on the Commission's own motion (Sections 15(1) and (3)). Such an investigation, in which the burden of proving discrimination would be upon the Commission's staff, would be a lengthy and expensive undertaking.

"The likelihood of disclosing actual cases of discriminatory rates and practices, of demonstrating their illegality, and of translating findings of unlawful discrimination into actual public benefits—in the form of lower petroleum products prices for the consumer—appears to be extremely small in relation to the effort involved in the kind of comprehensive investigation that would be necessary. It is the Commission's position, then, that we would not be justified in undertaking an investigation into possible discriminatory rates or practices of oil pipelines in the absence of much more positive information of potentially unlawful conduct than we have received heretofore."

At another place in his letter, Chairman Stafford had this to say:

". . . there is no evidence that the past or present practices of joint venture pipelines have discriminated against independent shippers."

Any pipeline company that fails to live up to its common carrier or other legal obligations should be investigated, prosecuted and punished. As Mr. Stafford pointed out, present laws are adequate for this purpose and there is no need for additional laws to provide such investigation and prosecution.

Oil pipeline companies must also comply with comprehensive safety regulations issued by DOT,[22] and with numerous state and federal environmental laws and regulations covering air, noise and water pollution[23] which include civil and criminal penalties for violation.

Adequate regulatory capability is provided in the Interstate Commerce Act, Elkins Act, Clayton Act and the Department of Justice Consent Decree. With this capability at hand there is no need to turn a highly efficient industry inside out on a recurring basis with charges and investigations on matters which have been covered previously in investigations and litigation.

We must also keep in mind that the entire oil pipeline system in the United States is available to all potential shippers, large or small, owners or non-owners, and is used by them. The following table covers oil pipeline shipments in the United States for the year 1973, broken down by non-owner shippers and shipper owners as well as between independents and majors. This should be conclusive evidence that rights of access to existing crude and products lines is available to all potential shippers. In addition to this, it should also be pointed out that independents participate in joint venture pipeline ownership with majors and independents ship with pipelines primarily owned by majors, and majors ship with independent pipelines.

On the subject of right of access to pipelines, the testimony of Mr. Robert E. Yancey, before this Committee on August 6, 1974, might create the impression that his company, Ashland, had difficulty in getting into the Colonial system as a shipper. This is definitely not the case. Ashland was invited to and attended a prospective shipper's meeting called by Colonial on May 7, 1963, attended by 23 prospective shippers (including nine Colonial owners). This meeting was held prior to the beginning of Colonial's operations and while the system was still under construction. Ashland's first request to become a shipper on Colonial was contained in a letter dated September 4, 1970 (acknowledged by letter on September 9), at which time we were advised what their initial requirements would be in December, 1970. In a letter dated October 21, 1970, Colonial urged Ashland to meet with our people as soon as possible in order to permit us to work their allocation

---

[22]18 U.S.C. Sec. 831 et seq. 49 CFR 195.50 et. seq.

[23]33 U.S.C. Sec. 407, 407(a), 411, 413.
42 U.S.C. Sec. 4321 et seq.
42 U.S.C. Sec. 4371 et seq.
33 U.S.C. Sec. 1161 et seq.
33 U.S.C. Sec. 1001-1015.

into our schedule in order to meet Ashland's time table. Ashland actually began shipping on the Colonial system on January 1, 1971. Space allocated to Ashland was in accordance with its forecast reduced by the percentage of proration in effect at that time under Colonial's proration policy which has been in effect and applied uniformly to all shippers since July, 1967, when the Colonial system first went on fulltime proration. Ashland's request to become a shipper was handled and built into our schedule as quickly as possible from a practical standpoint, conformed with the handling of all new shipper requests, and conformed to the time table which Ashland had set for themselves.

OIL PIPELINE SHIPMENTS IN THE UNITED STATES—CALENDAR YEAR 1973

	Product pipelines	Crude trunk lines	Gathering systems
Total number of shippers[1]	909	559	291
Number of nonowner shippers	806	491	252
Number of shipper owners	103	68	39
Owner shippers:			
Independent	17	17	9
Majors	86	51	30
Total	103	68	39
Breakdown of nonowner shippers and owner shippers between independent and majors[2] companies:			
Nonowner shippers:			
Independent	573	292	146
Majors[1]	233	199	106
Total	806	491	252

[1]Total for all pipelines. Many shippers ship through more than 1 pipeline system.

[2]A staff report of the Federal Trade Commission on its investigation of the petroleum industry, prepared at the request of the Chairman of the Senate Interior and Insular Affairs Committee and released by that committee in July, 1973, lists the following 18 companies as major oil companies: Exxon, Gulf, Standard (Ind.), Texaco, Shell, ARCO, Mobil, Socal, Sun, Union, Phillips, Continental, Cities Service, Getty, Standard (Ohio), Amerada Hess, Skelly, and Marathon.

The Subcommittee in its questioning of Robert E. Yancey and Richard C. Hulbert raised questions regarding pipeline earnings, return on equity, risk element, and dividends paid to shipper owners. I would like to review each one of these points as follows:

Pipeline earnings

Pipeline rates average about one-fifth (1/5) of that of railroads and one-twentyeighth (1/28) of that of trucks and can usually compete favorably with all marine transportation except ocean-going, long-haul, jumbo tankers. Moreover, the average level of pipeline rates has increased only 1.03 percent during the twenty-five year period from 1947-1972, in sharp contrast to the rates of rails and trucks which increased about 47 percent and 65 percent, respectively, during that period.[24]

The ICC has established earnings guidelines of eight percent for crude lines and ten percent for product lines as a fair rate of return on ICC valuation, [25] has issued extensive regulations requiring oil pipeline carriers to maintain accounts[26] and file valuation reports,[27] and has prescribed rules for depreciation and the publication of oil pipeline tariffs.[28] Valuations of pipeline properties have been kept since 1934 and are brought up to date annually.

An Elkins Act Consent decree covering 59 pipeline companies and their owners was entered in 1941.[29] This agreed judgment limits dividends payable to pipeline shipper-owners to seven percent of the latest ICC valuation. In 1957, the U.S. Supreme Court upheld the right of pipeline companies to earn money and pay dividends on properties purchased with borrowed funds.[30]

All but a few of the interstate oil pipelines are covered by this decree and comply with its provisions. Inasmuch as this decree is more restrictive than the 8 and 10 percent rates of return established by the ICC, Chairman Stafford of the ICC testified before the Select Committee on Small Business on June 14, 1972, that this consent decree "has had a tendency to keep pipeline carrier rates in line because the majority of pipeline carrier officials have, over the years, reduced their rates so as not to make a return on

---

[24]TAA Transportation Facts and Trends, 9th Edition, July, 1972, pp. 4, 7.

[25]23 ICC 115 (1940); 272 ICC 375 (1948); 243 ICC 589 (1941).

[26]49 CFR pt. 1204.

[27]49 CFR pt. 1260.

[28]205 ICC 33 (1934).

[29]U.S. vs. Atlantic Refining Co., et al.; Civil Action 14060; U.S. Dist. Ct. D.C. 12-23-1941.

[30]U.S. vs. Atlantic Refining Co., et al. 360 U.S. 19 3 L Ed 2d 1054; 79 S. Ct. 944-6-8-1959.

capital assets greatly exceeding the 7 percent of value allowed under the consent decree for dividend purposes."[31]

Although pipeline rates are not high, and have attracted about as many complaints from being too low as too high,[32] the ICC has adequate authority to deal with rate complaints as evidenced by the recent rate proceedings.[33] Notwithstanding this excellent rate comparison, and the existing regulations and litigation history, these charges of excessive earnings and high rates continue to be indiscriminately repeated.

Dr. Stewart C. Myers, Associate Professor of Finance of the Sloan School of Management, Massachusetts Institute of Technology, testified on February 20, 1974, before the Special Subcommittee on Integrated Oil Operations of the Senate Interior Committee, on profitability in the oil pipeline industry. His testimony, which was based on ICC statistics, showed that the industry weighted average ratio of net income to assets for all pipeline companies reporting to the ICC for the year 1971, a typical year, was 8.7 percent.

The ICC reports show after-tax income on a flow-through basis. When these income figures are adjusted on a normalized basis, the median income for all 97 pipelines reporting to the ICC in 1971 was 7.7 percent.

Return on equity

Reference is continually made to the so-called high return on equity of Colonial Pipeline Company, as well as other lines. As a true measure of profitability of a highly leveraged project, such a computation is absolutely meaningless. Forbes Magazine, an acknowledged authority on corporate profitability, states in its Annual Summary of National Industrial Profitability:

"The five year return on total capital is similar to return on equity, but is the profit expressed as a return on both stockholder investment and debt. The amount of debt a company incurs is based on a financial strategy of increasing return by using borrowed money, the interest on which is tax deductible. Sometimes,the strategy pays off fine, sometimes it backfires.

---

[31]House Subcommittee Hearings, Special Small Business Problems—Select Committee on Small Business, 92d Congress, Transcript p. 76.

[32]See testimony of Joseph A. Klausner, American Maritime Assn. 6/13/72 before Subcommittee on Special Small Business Problems; *Petroleum Rail Shippers v. Alton*, Note 31, *supra.*

[33]No. 35533 and 35533 (Sub.-No. 2), *Petroleum Products, Williams Bros. Pipe Line Co.;* No. 35533 (Sub.-No. 1), *Petroleum Products to Illinois, Iowa, and Missouri, Williams Bros. Pipe Line Co.;* No. 35540, *Petroleum Products, Louisiana and Texas to Midwest;* and No. 35720, *American Petrofina Co. of Texas, et al v. Williams Bros. Pipe Line Co.*

The degree of leverage and its consequences are so clearly the result of management decision-making that the return on capital is probably a more valid measurement of a company's profitability and its management's effectiveness."

The wide variation in debt-equity ratios and financing patterns makes the liquid pipeline industry's ratio of return computed on equity a meaningless statistic. The company stockholders under the so-called Through-Put Agreement are allocating a part of their debt capacity to the pipeline company and in doing so, are making a commitment of capital to the pipeline equivalent, in substance, to a direct equity contribution. In other words, the investment of the owners, under this financial commitment, is equivalent, as a practical matter, to the total capitalization of the company, not only for total investment costs but long-term operating costs as well. I have discussed this with a representative of one of the top investment banking houses in New York City, as well as representatives of a major public accounting firm, both familiar with pipeline financing arrangements, and they confirm this conclusion.

Colonial's return on total capital versus Forbes Magazine's computed nationwide industry median on the same basis for the last three years is as follows:

	1971		1972		1973	
	Col.	Forbes	Col.	Forbes	Co.	Forbes
Return on total capital:						
5-year average.....	7.97	9.0	7.02	8.5	7.13	8.2
Calendar year	7.07	6.8	7.08	7.3	7.03	9.0

Colonial has historically financed its long-term capital needs through 10 percent equity and 90 percent long-term borrowings. The security offered by the backing of its stockholders through the form of the so-called Through-Put Agreement has supplied Colonial with a ready market for its debt issues at favorable rates afforded by the high ratings assigned to its notes by the rating agencies. The ratings are the average of those of its stockholders. In other words, the rating agencies looked to the Through-Put Agreement and beyond that to the stockholders in assigning the rating to Colonial.

If there was no Through-Put Agreement on similar financial commitment by substantial companies, Colonial would not have been able to borrow money in the amounts required to meet its capital needs. This leaves the equity market as the only other means of furnishing equity capital, and investors would need a return commensurate with risk.

In order to induce investors to provide long-term capital to Colonial, a sufficient return (interest) must be offered the investor on debt issues, or in the case of the equity investor, a sufficient expected return on his invested capital must be offered to justify the risk over a secured debt issue. If Colonial were to raise capital wholly by equity in the public market place, its cost of capital would be much higher, and accordingly, Colonial would be required to increase its charges for transportation services.

The need for this increase is illustrated as follows: Colonial's rate of return for the last four years on total capital (90 percent debt + 10 percent equity) has been 7.1 percent. If Colonial were to use straight equity for financing, the cost of building its facilities would require a rate of return sufficient to attract investors commensurate with the risk involved. Uncertainties are great for Colonial and would be considered well above an average risk by an investor if no Through-Put Agreement from the owners existed. Shown below, based on 1973, is the amount of increase in Colonial's charge for transportation services which would be necessary to earn the investor three assumed rates of return on a 100 percent equity basis.

Rate of return	Tariff increase (percent)	Shippers Transportation cost increase 1 year
20 percent	86	$117,351,000
15 percent	53	71,866,000
12.8 percent (Forbes median)	38	51,853,000

Risk element

Again referring to Colonial, it has been asked, "Where is the risk involved?" and again reference is made to a throughput commitment at the very beginning of the project. I do not believe that there is a correct understanding involved here as to what is really meant by the Through-Put Agreement to which the Colonial stockholders are a party. These commitments are not just an agreement to ship X numbers of barrels of oil but, instead, commit the stockholders, and the stockholders only, not all shippers, to guarantee financing of the project totaling hundreds of millions of dollars by guaranteeing the payment of all cash-need shortfalls, including all operating costs and debt service to provide transportation service not only for shipper-owners, but also for the shippers who have no ownership interest and no commitment of risk at stake. These commitments typically run for 30 to 40 years. In the case of Colonial, there are nine shipper-owners whose names are on this commitment and 18 additional shippers whose names are not on this commitment. All 27 shippers ship products under the same published tariff.

Since the original Colonial system was built, Colonial has doubled its capacity at a cost of approximately $200,000,000. This additional $200,000,000 has been added to the original commitment to which the owners are a party while the non-owner shippers are able to utilize the availability of additional space without any financial commitments of any kind, other than the tariff obligations. Obviously, if the barrels continue to flow through the system and all cash requirements are met, risk is minimized. But over a period of 35-40 years, who can safely say this will happen? Consider the changes which have occurred just in the last year or so in the petroleum industry.

Colonial is currently making additional expansion studies for its system. Based on updated shipper forecasts, expenditures of $600 to $700 million will be required to accommodate these forecasts. But what about the imponderables which could affect the success of such a project? For example:

1. How will oil production on the Outer Continental Shelf of the Atlantic Ocean affect Colonial?

2. What about finished product imports to the East Coast?

3. Where will future refineries be built? New England? New Jersey? North Carolina?

4. How much oil will come from Alaska? Will its distribution pattern affect Colonial?

5. Will oil shale be a significant factor during the financial commitment period?

6. When and where will off-shore crude unloading facilities be built?

7. Will there be enough electric generative capacity available to support a major Colonial expansion? What will the power costs be?

8. What will happen to future refined product demands? Recession? Depression?

9. Will the capital requirements be available for borrowing?

10. What will be the extent and effect of Governmental regulations and changes in political climates?

Risks? Certainly there are tremendous risks in underwriting projects of this nature and certainly those assuming these risks are entitled to a reasonable return on the total invested capital. It would be naive to expect oil companies or any others to enter into similar long-range financial commitments without having some part of the ownership and reward for risk assumption. It must be kept in mind that common carrier pipeline transportation differs from other forms of common carrier transportation in that pipelines serve only one industry, moving over fixed routes in one direction and with fixed points of delivery. Other modes of transportation can handle various types of freight for many different industries flowing in different directions. Once a pipeline is installed, it is dependent upon the oil industry it serves to have liquid petroleum to move over these fixed routes. While it is true that in some instances line reversals might be practical, such would completely change the economics of a project.

While on the subject of risks, I would like to point out that in the case of Colonial, its fixed costs for 1974 were approximately $80,000,000. This means that for every day the system is completely shut down for whatever reason, the company is incurring a cost of approximately $219,000 with zero operating revenue to apply against this cost. Fortunately, partial or complete shutdowns do not occur frequently. However, except for planned maintenance or construction, shutdowns are unpredictable, but do occur and when a system is operating at full capacity there is no way to make up the loss in revenue.

Dividends paid to shipper-owners

In the earlier hearings it was asked, "Is it not true that in the case of a shipper-owner line such as Colonial, that there are substantial dividends paid which some might suggest are rebates?" The earnings and dividends of interstate oil pipelines have been the subject of extensive litigation resulting in the Elkins Act Consent Decree. This decree limits pipeline annual dividend payments to shipper-owners to no more than seven percent of its ICC valuation, which valuation includes properties purchased with borrowed funds. Colonial operates within the framework of this decree, and reports annually to the Justice Department on such earnings and dividend payments.

A seven precent maximum return in the form of dividends can hardly be called excessive and, by today's standards, is hardly "substantial." Mr. Yancey of Ashland does not think pipeline earnings are excessive or adequate since, in his testimony on August 6, he stated, "We would much rather pay the tariff on that basis and put our money into other areas where we get a much higher rate of return." And further, in referring to future expansion of Capline, a joint venture project in which Ashland is a participant, he stated, "The investment necessary to put that thing in (expansion) we would hope somebody would put it in and charge us the tariff to go through it so we could use our capital for something else." Colonial had done this for its shippers, doubling the capacity of its line since the original construction.

Divorcement

Senator Floyd K. Haskell in his testimony before this Committee on August 9, 1974, stated:

"I think the fact that the ICC is ineffective is probably borne out by some rather negative evidence. It is my understanding that until last year, no complaint was filed against an oil pipe line with the I.C.C. . . . I would suggest that in commercial operations not to have a dispute between a carrier and a customer is so unusual that it rather indicates that the economic advantage of the pipeline owner is such that a potential customer, if he wants to get his product to market, does not dare sue that pipeline. I think this reasonably self-evident."

In the first place, Senator Haskell's understanding that no complaints were filed with the ICC until last year is not correct according to testimony of the Chairman of the ICC before the House Select Committee on Small Business, in June, 1972. But more important, surely Senator Haskell does not think that all disputes between a carrier and a customer are carried to the ICC. In the pipeline industry, as in all industry, disputes with customers occur on various matters but almost all of these disputes are worked out before any "last resort" legal steps have to be taken. Testimony given before this Committee by Charles P. Siess, President of APCO Pipe Line, Inc., is an example of this.

According to Mr. Siess' testimony APCO had a dispute with Sun Pipe Line Company, which apparently was satisfactorily worked out before any complaint was filed with the ICC. Of course, disputes occur between pipeline carriers and their shippers. But, in practically all cases, these problems are worked out in the normal course of business to the mutual satisfaction of the parties. I believe, as I stated in testimony before the House Small Business Committee in June, 1972, that the statement of some pipeline industry critics to the effect that shippers do not dare to sue a pipeline is a myth and is only used because these critics do not have any tangible evidence to support their position. At the same time, they refuse to take into consideration testimony already on the record which contradicts their criticisms.

Congress has considered the divorcement issue several times and has concluded that such action would be unwise. In 1906 Congress had to cope with the problem that certain railroads were refusing to give switch connections and car service to independent coal mine owners. At that time the transportation laws were not fully developed and the ICC lacked authority to handle the situation. Congress corrected this problem by enacting a law making it illegal for a railroad to transport any product (other than timber) in which it had an interest. This has become known as the "commodities clause."[34] In this same legislation, oil pipelines were brought under the Interstate Commerce Act.[35] Congress considered, but wisely decided against including pipelines under the commodities clause. During these same hearings consideration was given to extending the clause to all common carriers by an amendment to the Elkins Act, but this was also rejected.[36] The commodities clause has never been extended to any other type of carrier.

Since 1906 additional teeth have been added to the Interstate Commerce Act to prevent discrimination as to rates and service which would have allowed the ICC to correct the coal problem without the commodities clause.[37] The Elkins Act also prohibits concessions, discrimination or rebate.[38] There is really no need for the commodities clause today and it should be repealed. The Transportation Association of America has so recommended. To prevent any mode of transport from hauling products for anyone is contrary to the common carrier concept, which makes service available to everyone on the same terms and conditions.

Divorcement bills have been introduced in Congress sporadically since 1906, but have never been seriously considered by Congress because of the almost total lack of complaints from pipeline shippers.[39] In the 1957 hearings before the Senate Subcommittee on Antitrust and Monopoly, the Independent Producers Association of America rose to the support of oil pipelines on the divorcement issue.[40]

Dr. Walter M. W. Splawn, Special Counsel to the House Committee on Interstate and Foreign Commerce, prepared a "Report on Pipelines"[41] in 1933 which states the case against divorcement very well:

"Oil pipelines are found as a result of this investigation to be plant facilities in an integrated industry. They are very different from railroads in that railroads carry all manner of freight whereas oil pipe lines are limited to one product: Petroleum carried in one direction, from a diminishing source of supply. Pipelines have been built primarily by oil companies. It appears

[34] 49 U.S.C. Sec. 1 (8).
[35] 49 U.S.C. Sec. 1(1) (b).
[36] 40 Cong. Record 6455 (1906).
[37] 49 U.S.C. Sec. 1 (12), (13); Sec. 10.
[38] 49 U.S.C. Sec. 41 *et seq.*
[39] Johnson Petroleum Pipelines and Public Policy; Wolbert, American Pipelines, p. 3.
[40] *Ibid.* at 437.
[41] H.R. Report No. 2192, Parts I and II, 72nd Congress, 2nd Session (1933).

very difficult to apply the 'commodities clause' to oil pipelines. If oil companies were forced to sell the pipeline companies, who would buy them and who would build to newly discovered oil fields? It appears that whatever regulation of oil pipelines may be necessary, it may be provided in recognition of the character of pipeline transportation and its relation to the oil business."

Divorcement of pipelines from their owners would achieve no useful purpose, but would deprive this country of needed pipeline service and contribute to the energy shortage. The consumption of petroleum products in this country has grown sharply over the last decade. Unfortunately, domestic oil exploration has not kept pace with demand and crude reserves are declining.

If this shortfall in petroleum energy is to be met, new pipelines must be built from new oil fields or seaports to refineries and from refineries to new or expanding markets. The use of oil pipelines for oil transport is increasing faster than the demand for petroleum products. This is particularly true for products pipelines. The total domestic demand for refined products increased 51 percent from 1960 to 1970 but the use of pipelines for transporting such products increased over 144 percent in the same decade.

There has always been a tremendous logistic problem in the distribution of petroleum products in the United States. For example, the 17 Eastern Seaboard states in PAD District 1 consume 39.8 percent of petroleum products, but produce only one-half of one percent of the crude oil produced in this country. East Coast refineries process only about one-fourth of the demand for the area. PAD District 2, which includes 15 midwestern and mid-continent states, also depends largely upon pipeline transport from other areas. This complex logistic problem has been compounded by the present energy shortage, the uncertainty as to volumes, locations and kinds of imports and ecological impediments.

The oil industry has always included pipelines in their long range planning and if the pipeline industry is permitted to function as it does today, there can be no doubt that the required pipeline space will be constructed as needed.

If this country should make the serious mistake of experimenting with pipeline economics through divorcement, it is very doubtful that the necessary pipeline space would be built to accommodate the ever increasing demand for petroleum, even when the supply problem is solved. The continued operation of some now in existence would be in doubt, particularly the marginal lines serving depleted oil fields or stripper wells, or refineries with reduced through-puts.

Certainly, the cost of financing pipelines would sharply increase, since any new project would have to compete on the market generally for speculative money and the pipeline projects could no longer take advantage of the low interest rates and long term private financing made possible by financial commitments and the shipper-owners, high credit ratings.

Pipeline rates would rise because of the increased costs of money and possible reduced through-puts by former owners who have alternate transportation sources. After divorcement, the seven percent dividend restriction in the Elkins Act Consent Decree would become a nullity.

The financial stability of pipeline companies could also be jeopardized by doubts as to the binding effect of financial guarantees after divorcement.

There is a serious question as to the constitutional validity of a statute which would take valuable pipeline properties, without adequate compensation, from their owners who had risked their capital and credit to create the successful ventures in full compliance with existing laws. At any event, such divestiture would undermine the confidence of the oil and pipeline industries, banking institutions and general public in the integrity of the Federal government and its elected representatives. Certainly, such action would be contrary to our American philosophy of fair play.

Divorcement would have no beneficial impact on the competitive environment. The location of the lines and connections to origin and delivery facilities and the list of shippers would remain unchanged. Pipelines cannot be picked up and moved to new locations.

One of the hazards of building a crude line is that the volume to be transported from a producing oil field diminishes as the field is depleted. Many of the great oil fields have now reached the stripper stage where the operation of gathering or trunk lines has become marginal and some pipeline service would undoubtedly be abandoned if the pipeline owners have no interest in the crude production when the service to be provided in uneconomical from the carrier standpoint. It is significant that divorcement legislation has received no support from independent producers. They fear that divorcement would deprive them of the present advantages of an assured market and pipeline transportation.

Pipeline carriers unaffiliated with a producer or a refinery are concerned only with transportation for profit. Their managements have a legal obligation to refuse extensions or continue service unless economically justified. Gathering oil from a flush field is much cheaper per barrel than from a field of stripper wells. In many cases the marginal wells are kept pumping and the pipelines operating solely because of the needs of the pipeline owner for the production. Today, it is essential that we squeeze the last drop of production from our marginal wells and encourage secondary recovery. If pipeline rates are raised proportionate to the costs of service in depleted fields, the oil could be priced out of the market and the production lost. The premature shut-down of "stripper wells" would lose this production forever due to the prohibitive expense of re-pressuring the reservoirs at a later date. As long as the reservoir pressure is maintained stripper wells can be pumped at a reasonable cost and maximum recovery obtained.

Mr. Ray M. Johnson, speaking for the Oklahoma Stripper Well Association in 1939 made the following statement as to the consequences of divorcement, which is still valid today:

"For the whole State of Oklahoma, it has been estimated that if the present stripper wells can be continued to be operated until water flooding is justified, these old, hoary petroleum relics will yet produce an ultimate recovery of four and a half billions of barrels of oil equalling the State's entire past production.

* * * * * * *

"Under conditions as they now exist in the oil industry, a great majority of all the Nation's oil wells are served by pipeline companies where one affiliated company purchases the oil and the oil is delivered to another affiliate. These affiliated companies realize that in the cycle of events these temporary periods of too much oil are always followed by periods of too little. They have enormous investments in refineries and marketing outlets to protect, and are eternally concerned with the permanence of supply and how to keep their pipelines filled and their refineries and filling stations running.

"Pipeline service has in fact been maintained in many cases after the gross income from it has ceased to pay operating expenses and the price paid for the oil has been the same, quality considered, over wide areas regardless of differences in the cost of gathering it. These practices which have developed in the industry have won the approval of the producers as a whole and in consequence among the thousands of small producers the proposal to disrupt the present relationships of pipelines has found no substantial support."[42]

The problem of divesting a single entity is immense.[43] Dismembering a multibillion dollar pipeline industry composed of many pipeline owners and literally thousands of individual systems joined together by multiple interconnections, joint tariffs and joint ventures, and financed by thousands of individual note holders, would be such a mammoth undertaking as to be

---

[42]T.N.E.C. Hearings Oct. 4, 1939 at 1130.
[43]Congressional Record 7-16-73 S. 13598.

practically impossible. Our already crowded court system is simply not equal to the task. Certainly, no such divestiture could be accomplished during this generation, and perhaps not before oil is replaced by other energy sources.

Even the most optimistic theorist could not expect to sell billions of dollars worth of pipelines by private negotiated sales. Sales of stock by public offerings would have little appeal to investors at the present regulated level of earnings. Utility stocks on all the exchanges have been depressed for years even though they have monopoly franchises and rates of return based on a cost of service basis.

It is unbelievable that serious consideration could be given to disrupting an industry, including both large and small operators, that has served the nation so well for so many years, at a time when pipelines are needed most.

Even if all of these divestitures could conceivably be accomplished, what would it mean to shippers and consumers? It would mean:

1. There would be a tremendous disruption of a vital and nationwide transportation system.

2. Existing lines would continue to be located where they are. It would be impractical and really unnecessary to change the physical format of the present nationwide system of pipelines.

3. Service to shippers would not improve since currently the most efficient form of transportation known for the movement of petroleum liquids is being provided on a non-discriminatory basis.

4. Costs to shippers would increase since financing costs to carriers would be substantially higher, and the seven percent earnings limitation for dividends would no longer prevail.

5. Much needed expansion of existing pipeline systems and construction of new systems would be slowed down substantially because of funding problems and the difficulty of private interests in getting the necessary support to justify the huge investments involved.

Consider the liquid pipeline industry in the light of the following:

1. The least costly form of transportation to the shipper by far for overland movements of freight. It costs eight times more to mail a postal card from Houston to New York than it does to ship a gallon of gasoline over the same distance.

2. Pipeline rates generally have remained constant for 25 years. This is due to a large extent to the economies of scale and innovative ideas and techniques.

3. No serious environmental problem. The industry's environmental track record is very good.

4. Liquid pipelines move over 22 percent of all inter-city freight (including mail) at less than one and one-half percent of the nation's total freight bill.

5. The liquid pipeline industry has 12 percent of the total investment of all regulated carriers, with only 4.2 percent of the operating revenues of such carriers. It has half again as much investment as regulated trucking with about nine percent of the trucking industry's revenues.

6. All of the above is done with no government subsidies or loans, Federal, state or local.

It is to be assumed that the purpose of these hearings is to find out what the facts really are within the various areas of investigation of the Subcommittee. Congress has a responsibility to be objective in reaching its conclusions that are to be translated into law, and it can only be objective if it takes into consideration fully what is actually happening. The facts are available to this Subcommittee and it is not overly difficult for this Subcommittee to obtain these facts. The pipeline industry stands ready to continue to cooperate in making the necessary information available, or to assist in confirming the situation that actually exists. The interest of the public will not be served if this Subcommittee fails to pursue this investigation in this fashion.

EXHIBIT 1

COLONIAL PIPELINE COMPANY

Proration Policy

Colonial's main line and some of its stub lines become completely full from time to time and are prorated among all shippers on an equitable basis. New shippers who are accepted during times of proration must realize that space allocations for all shippers may be less than nominations, depending upon the extent of proration in effect at the time. Proration is handled as follows.

Allocations to new shippers

A "new shipper" is any shipper (as defined in Colonial's published tariff) who does not qualify as a regular shipper. The shipper will become a regular shipper at the end of 13 months from the beginning of the first month in which the new shipper received deliveries.

A new shipper requesting space during periods of proration (and who has otherwise satisfied applicable requirements of the tariff rules) will be allocated space as follows: New shipper forecast volumes will be added to the aggregate of the regular shipper forecasts to determine the demand for pipeline space in each month. The forecast volumes of each new shipper will then be reduced proportionately by the extent that total forecast volumes exceed total line capacity. The total forecast volumes of regular shippers will also be reduced by the same percentage.

Allocations to regular shippers

Space (adjusted as described in paragraph 2.5.1, will be allocated to regular shippers on the historical percentage basis that the total barrels shipped by and delivered to each regular shipper bears to the total barrels shipped by and delivered to all regular shippers during the 12-month period ending one month prior to the month of allocation.

General requirements

Colonial will carefully examine forecasts using every means available to ensure that they are true and realistic and will challenge any forecasts which appear to be inflated. To maintain equitable allocation of space on the system, if a new shipper is unable to deliver products equal to the space allocated to it and such inability has not been caused by force majeure or other cause beyond control of such shipper, satisfactory to Colonial, its volumes for the following month will be reduced by the amount of allocated throughput not utilized during the preceding month.

In no event will any portion of an allocation granted to a new shipper be used in such manner that it will increase the allocation of another shipper beyond what it is entitled to under the proration policy. Colonial will require written assurances from responsible officials of shippers respecting use of allocated space stating that this requirement has not been violated. In the event any new shipper should, by any device, scheme or arrangement whatsoever, make available to another shipper, or in the event any shipper should receive and use any space from a new shipper through violation of this requirement, the allocated space for both shippers will be reduced to the extent of the excess space so made available or used in the shipping cycles next following discovery of the violation which are under proration.

Colonial reserves the right to adjust any shipper's allocation in the event of a substantial sale, transfer, or change affecting that shipper's throughput which served as the basis for the allocation.

If any shipper, new or regular, is unable to deliver products equal to the space allocated because of a strike, labor slowdown, or fire or other force majeure, Colonial may adjust the allocations to prevent the loss of that shipper's allocation percentage.

From: Colonial Pipeline Company manual titled "Shipping Instructions."

11

The Southern Caucus Testifies

TESTIMONY OF J.W. ADAMS III, C.R. JACKSON, PAT GREEN AND J.R. JOHNSON BEFORE THE SUBCOMMITTEE ON ANTITRUST AND MONOPOLY, COMMITTEE ON THE JUDICIARY, UNITED STATES SENATE—JANUARY 29, 1976

Summary: Under divestiture, there would be no assurance of adequate supply at wholesale or retail or that supplying companies would continue to compete to supply the various stages of operations. Divestiture would mean higher prices because there would be less competition. Also, supply to distributors in rural markets might be unprofitable and therefore in jeopardy; there would be no credit card program, no financing, and no credit terms such as are now extended by suppliers. Economies now available from crude production all the way through marketing would be lost.

J. W. Adams III speaks . . .

Thank you, Mr. Chairman. My name is Billy Adams, and I am a gasoline jobber from Macon, Georgia. As one of 15,000 jobbers nationwide, I am grateful for the opportunity to express some of the feelings by fellow jobbers and I have about this proposed legislation.

First, let me say something about the jobber's function in the market place. Many people who buy name-brand gasoline at a service station are under the impression they are buying from the company's dealer directly. Actually, most gasoline outlets across the country are served by jobbers, and not by the major oil companies.

Whether we market under a brand name or as unbranded gasoline, we are the people who most often supply the capital and the knowledge of local market conditions—responding to customer's needs on the level that meets their demands. When a new piece of property is sold or zoned for residential development, it is not the large oil companies who hear about it in their distant headquarters. It is not the large oil companies that determine the most promising locations for new service stations. Nor is it the large oil companies that put up the risk capital and open stations before the first residents move in. This is primarily a function of the jobber, who is on the scene and who is an integral part of the community in which he serves. For this reason, prior to FEA, the major oil companies constantly competed for supply relationships with aggressive, innovative jobbers.

This system has resulted in competition among jobbers to supply consumers with the best service at the lowest price, and competition among major oil companies to supply the best jobbers with their product. However, recent government intervention in the oil industry has caused an erosion of this competition, and in my judgment, any effort now to break up the supplying companies will result in even less competition, with the greatest punishment being inflicted ultimately upon the American consumer.

My own operation consists of 20 outlets, 14 of which are full-scale service stations and 6 of which sell only gasoline. I am an independent jobber, which means I may contract with any supplier I wish to, but I have chosen to contract with a major oil company marketing under a nationally recognized brand name. This has enabled me to take advantage of my supplying company's national advertising and nationally-honored credit card service.

On the other hand, some of my jobber competitors in the Macon area have chosen to build their business by marketing an unbranded product for cash at a lower price. Between the two approaches, consumers in the Macon area are given a full range of choices.

My particular brand is Amoco. That affiliation, however, doesn't restrain my freedom to deal with other companies. In addition to my Amoco outlets I own interest in a full-service carwash, which represents an investment of about $600,000. It so happened that Mobil gave us a better deal on gasoline prices than Amoco, so that particular operation flies the Mobil flag.

The point I make is this: In my operations (and I've got a pretty typical jobber operation) I have the freedom to sell unbranded or branded gasoline, and until FEA came into being, I had the freedom to choose my own supplier.

Now, on the heels of the massive problems created by FEA, comes this further government proposal to divest major oil companies of certain of their functions. Its proponents claim this will stimulate competition in the market place. I fear it will do just the opposite.

Under the present structure, as a jobber I have some stability assured for my business. I can be relatively assured of my supply for periods of up to five years on renewable contract. On the basis of that assurance, the banks can risk money on me, and in return I can risk money on service station locations, enabling me to offer reasonable stability in employment for those who work with me. I know I can buy somewhere in the neighborhood of 15 million gallons of fuel per year and that I can sell that much making a reasonable profit for the risks taken and the money invested.

If a divestiture bill passes, I will no longer have the assurance of any of these things. If marketing is divorced from refining, I would have no assurance that my suppliers would have adequate product to supply my operation. If refining were divorced from production, I would have no assurance that refining and distribution would have access to crude production on a stable basis. Even if I could get sufficient gasoline for my customers, there is no assurance that they would be able to pay with credit cards as before. My relationship with my customers would face drastic changes. There would also be no assurance that supplying companies would continue to have the incentive to bid against each other for the privilege of supplying my operation.

At the present time, my fellow jobbers and I represent part of the business community in Macon, Milledgeville, Gray and Forsyth, Georgia. We are a part of the local flow of funds. We all have employees, we all have loans with the bank, we all supply services that make life in our area efficient and comfortable for the consumer. In short, we enjoy a stability and an orderliness in the marketing of gasoline in the Macon area.

In the face of our current economic problems, it seems to me Congress should be particularly hesitant about tampering with a system that has served us so well for so long. The time is just not right to experiment, or to even consider the implementation of any measure that has the potential of breaking up our present stability and orderliness, or that might precipitate a major disruption of the market place. I my opinion, divestiture would do precisely that and, frankly, I fear the damaging consequences that would be felt in the Macon area.

It would appear the primary purpose behind this proposal is to use the force of government to punish an industry that, until the O.P.E.C. nations raised the price of oil, was more distinguished for its low prices and gas wars than for anything else. If the conclusion is that the major oil companies are responsible for the energy crisis, then the conclusion logically follows that the free market system is responsible for the energy crisis. I simply do not believe that to be the case, and I have to believe an overwhelming majority of the American people share that opinion, preferring instead to let competition in the market place solve our energy dilemma. After all, this is the system that has been responsible for providing us with the highest standard of living the world has ever known. Using government to break up the supplying oil companies of this nation is not consistent with the principles underlying that system. For this reason, in my judgment, the passage of divestiture legislation would only contribute to a lowering of our standard of living.

Finally, may I respectfully suggest that if Congress really wants to accomplish something constructive to help provide my customers with plenty of product, good service, and low prices, then return capital incentives to the market place by, first, eliminating unnecessary bureaucratic intervention in the oil industry, and second, by putting the brakes on inflation. A sound dollar coupled with freedom in the market place will not only solve our energy problems but will keep our nation strong and solvent.

Thank you, Mr. Chairman.

Charles R. Jackson speaks . . .

Mr. Chairman, I am Charles Jackson, President of Jackson Oil Company, Incorporated, a small branded jobbership founded by my father in 1936. Jackson Oil Company is an independent branded distributor (jobber). We have been in this business nearly forty years. Our company employs eleven people who have families averaging 3.8 per household. We are proud that our little company, through meaningful and productive employment, can support eleven families with income, health and life insurance and retirement benefits.

I appreciate the opportunity to testify here today, to attempt to express our views on S. 2387 and the issue of vertical divestiture of the petroleum industry. In particular, I hope to convey our grave concern about the impact of such legislation on our business, the impact on the business of our customers, and the impact on the consumer in our market.

We market principally in Chesterfield County, South Carolina, a rural county of 793 square miles and 34,000 in population. You can see that we are very rural.

Our business is principally gasoline, most of which is sold to retail gasoline dealers who in turn sell to the retail consumer at the gasoline pump. We also sell gasoline and diesel fuel directly to farmers, heating oils directly to the home heating customers, and lubricating oils and greases to industrial customers.

During the forty years we have been in business, we have managed to grow steadily—improving operations and profits, and today we have a small but sound company. We believe we are an asset to our community. Over the years we have vigorously participated in and supported the civic and political affairs of our town and county, and I hope we have made and will continue to make some lasting contributions to the social and economic development of this area. If so, we must give a large part of the credit to the economic viability of our business.

We choose to operate our business as a major branded distributor. Our supplier is EXXON from whom we buy under contract, the terms of which are standard buy-sell contract provisions. You hear charges today that the major oil companies plan to take the jobbers over and put the jobbers out of business or that the majors really control the jobbers. We don't believe that is true, for the reason that the majors cannot serve the rural markets. It is simply too expensive for them to do so. Also, they know we can do a better job of serving the consumer in our market. We know the people, the customers. The major knows the best way to serve this market is through the jobbers.

We haven't asked our supplier to do our work for us, or to eliminate business risk for us. What we do expect is a supply of a quality product at competitive prices and certain technical assistance and service. Our supplier, on the other hand, expects us to promote and represent its brand and trademark and operate our business in such a way that its brand image is enhanced.

I see nothing coercive or sinister about this kind of relationship. Any sort of productive relationship, business or otherwise, is a two-way street requiring confidence, integrity, the ability to perform, and a willingness to negotiate points of difference. We have never expected our supplier or the government to protect us from failure; rather, we are willing to take our risk with the discipline of the free marketplace.

Rather than our supplier being a detriment to our growth, we feel we have enjoyed a number of advantages as a result of associating ourselves with a major branded product.

1. We have operated with the assurance of a supply of quality products at competitive prices.
2. We have benefited from a national advertising program and sales promotional activities resulting in immediate identification and acceptance by the consumer.
3. We have had a national retail credit card program to offer our customers.
4. Our supplier has provided financing for expansion projects and credit terms for our product purchases.
5. Our supplier has provided us assistance in business counselling and assistance in developing some of our management techniques.
6. Our supplier has provided us with professional technical assistance in the area of engineering, planning and surveys of oil and lubrication requirements for some of our industrial customers.

We have a lot of years, a lot of hard work, and a lot of our investment dollars in associating our supplier's brand name with our sales, our services and our company image. It is vital to us and to most of our customers that this relationship with our supplier continue.

Each of our retail dealer customers owns and operates his business as a small independent family enterprise. Many of them own the land and building where they operate.

Twelve percent of our sales are direct sales to farmers. In all but one case, these people are small farmers—some farming less than one hundred acres of land. Most of these farm customers work at regular jobs—usually in an industrial plant—but the part-time farming activity is important to them. About eight percent of our sales volume is to 500 home heating oil customers scattered about our county who expect us to supply their home heating requirements in an efficient manner and at competitive prices.

Why are we concerned about the impact of S. 2387 on our company's business, the business of our principal customers, and the consumer? We believe that such legislation may gravely jeopardize the economic viability of our small company and the livelihood of its employees. We believe such legislation may do irreparable harm to the business of our principal customers, the small retail dealers who sell to the consumer at the pump. Further, should such legislation be enacted, we are convinced that the consumer, served by our retail dealer customer, and the farmers and home heating oil consumers served directly by us, would all bear the burden of higher prices, inefficient and inferior service and questionable supply of their petroleum requirements.

First, regarding the impact on our company's business, the question arises: Given divestiture, who is going to supply us?

Supplying distributors like those of us in rural markets may be unprofitable, and may not fit into the scheme of the new marketing companies. We can't seem to get an answer to this question. I think the answer is, at the least, that our supply would be in jeopardy. We also believe our brand identification, which we value so highly, may be lost. We may have no national advertising program to support us. We may have no credit card program to offer our customers. The financing and credit terms presently extended by our supplier may not be available to us. The technical assistance provided us at no cost by our supplier may be lost. Furthermore, we believe our cost of goods would increase at least by the amount of the cost savings that vertical integration provides. Product costs to the distributor are bound to increase. There is no way to avoid this.

In addition, if some suppliers do withdraw from the rural market, we see the potential for giant chain operators to enter our market and others like it by taking over available retail properties. If this happens, the result will be fewer companies — not more companies — in the marketplace. As these new, larger retail marketing companies look for growth, they could force small distributors, such as my company, out of business. Our small branded independent jobbership will lose with divestiture.

Second, regarding the impact of divestiture on our principal customers, the thirty-six retail dealers mentioned previously, we believe, given divestiture, their small independent, family-owned businesses may be in great jeopardy. If Jackson loses his supply, who is to supply Jackson's dealers? If Jackson loses the benefits of brand identification, so do Jackson's retail dealer customers.

Many of these dealers have very small storage, varied credit needs and buying habits, and are widely scattered over our county. The new retail marketing companies, being highly profit-oriented, will be interested in the "cream" of the marketplace. The "cream" will not be the Mom-and-Pops and corner service stations of rural Chesterfield County. Our retail dealer customers will lose with divestiture.

Finally, what of the impact of divestiture on the consumer? During the embargo and the horrendous experience of the gasoline lines, we were reminded of something. We learned that the retail consumer in Chesterfield County, or any rural county, is totally dependent upon the automobile. We have no alternate transportation. You drive your car or ride with a neighbor or you don't go.

I won't go into the details of the hardship of the embargo on those of us in rural areas. We did learn that most of our people who work in our industrial plants drive twenty or more miles per day to work and back home. On a per capita basis, we use a lot of gasoline in Chesterfield County. The impact of divestiture on the consumer will be one of less competition in the marketplace, fewer places to buy gasoline, increased prices for gasoline; and all of this to contend with when one is totally dependent upon the automobile for his livelihood.

Regarding the farmers and home heating consumers we serve directly, given divestiture much of the above applies to them. Who is going to deliver to the small farmer who has a 280 gallon tank and wants to pay his bill once a year? Who is going to deliver to the heating oil customer who refuses to buy more than a 100 gallon storage tank and probably can't afford to? We have many such examples. Would the new marketing companies, with stockholders demanding marketing profits, be sensitive to the needs of these customers? I think not.

The consumer will lose with divestiture.

In summary, I feel our jobbership and thousands like it will lose by divestiture, our retail dealer customers will lose with divestiture, and the consumer will surely lose. As a citizen of this great country of ours, as a consumer, as a voter, as a taxpayer, as a holder of public office I ask these questions — Who is to gain by this dismemberment of the oil industry? What is to be gained from this most serious proposal of divestiture? What is the economic gain? What is the social gain? Who is to benefit?

I can find no benefit—no gain—and some potential grave consequences. Thank you for this opportunity to present our views of this question.

Pat Green, Jr. speaks . . .

I am an independent petroleum jobber serving several rural counties in Mississippi, and domiciled in Collins, Mississippi.

I am very much opposed to the idea of divestiture in the petroleum industry of this country. The largest companies in our industry are being attacked because of their size and are charged with being non-competitive and insensitive to consumer welfare.

In my remarks today I will comment mainly on the facts as I know them of competition in the marketplace and its ultimate benefit to the consumer because marketing is my business. However, I would make one observation about the allegation that these companies are too large, and that is simply this—they *have* to be. The amount of capital required to explore for oil worldwide, then to refine the crude once it is found, involves the commitment of billions of dollars. If the bottom line of a company's financial statement reads in the hundreds of millions I ask you to measure this large amount of money in terms of return on investment and you will see that their return is lower than many industries in this country and not excessive.

I function in the marketplace as a petroleum jobber in the following manner. I buy petroleum products from a refiner, Phillips Petroleum Company primarily. I transport that product from a terminal to a retail outlet that may be owned by me or by an individual dealer. It is then sold as branded and, in some cases, as unbranded gasoline depending on market needs. My margin, or profit, is derived from my ability to perform this function and sell the product at a price greater than my cost. If there were no competition in our industry my job would be easy. In fact, if Congress would legislate everyone a place in this industry, the job of transporting and operations would be simple.

If there is no competition in our business, then I have wasted some long hard hours and capital funds attempting to woo new customers and at the same time keep the ones I have. Mr. Chairman, the American consumer, my customer, needs no Federal protection. All he needs is the freedom to do business where he chooses, and there are no less than 15,000 independent oil jobbers and 200,000 retail station dealers attempting to convince him that theirs is the station to patronize.

A classic example of this free enterprise at work can be seen now. In the last few years the concept of self-service at gasoline outlets has gained in favor with the consumer. Under this mode of operation, an inexpensive building can be installed and savings in labor and investment are passed on as lower pump prices. This innovative method of selling came from independent marketers such as myself. If the larger integrated companies had the control in this industry that is purported, self-service would not exist, for it is mainly the large oil company that has the investment in properties costing $100,000 or more.

These high-cost stations are the ones suffering large gallonage losses to independent wholesalers and retailers, branded and unbranded. The point I hope you'll bear in mind is that the consumer is making this new marketing concept work. His interests are being met not because of legislative edict but because there are lots of us out there who want his business, with each trying to outdo the other in terms of price or service to obtain his business. And I can honestly say to you, my concern is not competition from direct major operations but from fellow independents who have historically been the innovators in marketing.

But what happens to my business if divorcement and or divestiture is ordered? *Frankly I'm not sure.* I do know this. The business of marketing petroleum products could then be dominated by large national chain operators who could exert more control in the marketplace than the so-called majors do now. In fact, there are several in existence now with the geographic diversity and available financial resources to do just that.

On the other hand, you now have a marketing system for petroleum products that has been developed over many decades. An everchanging system, yes; but the changes are not being brought about by a few large companies manipulating markets rather by our customers and our competitors, and the end result is the sale of product at the pump by the most efficient means. The proposition of divorcement is of such magnitude that no one can be sure of its consequences. It would end business arrangements that have been in place for years, and replace them with what? Who would supply the product I need for my customers, and on what terms? And at what price? Where would the capital come from to acquire transportation equipment, storage facilities, and the support services?

The business of furnishing this nation its energy needs is not a simple process, Mr. Chairman. If this legislation is passed by Congress, you will dismantle a system that has evolved over many years and has performed on the whole very well. In its place would have to come a bureaucratic maze that would make Congress yearn for simple problems like efficient delivery of the mail and profitable railroad operations. I don't believe, Mr. Chairman, that the advocates of this Bill realize the magnitude of their proposals. I do believe, however, that they know such legislation would take out of the free enterprise category one of our leading industries—an industry that has given the American people bargain-basement energy prices when compared to other industrialized nations. If these companies are broken up, the technology, organization, refining capacity, and high risk capital for exploration will never again be in private hands as a highly efficient business organization. And even though the Federal Government may provide the needed capital it will never match the present efficient use of this capital. And who will pay the difference? The consumers, of course, who are the voters and taxpayers of this country.

I would urge you, Mr. Chairman, to proceed very slowly and deliberately in considering legislation of the magnitude. Surely some other remedy is available if it can be proved that the large companies of this industry are using their strength not in the public interest. Why not use existing laws that are now in place to deal with monopoly and noncompetitive behavior if it exists?

Finally, Mr. Chairman, it's imperative that you and your colleagues, in your positions of high public trust, use the perspective toward world affairs that is available to you here in Washington to view other countries, Great Britain for instance, and see where excess government involvement in the economy has led them. Hopefully it will be your conclusion that our own economy needs less not more Federal involvement. Thank you, Mr. Chairman.

John R. Johnson speaks . . .

I am John R. Johnson, President of Johnson Oil Company, Inc. of Morristown, Tennessee, a jobber, or wholesaler, of gasoline, diesel fuel, kerosene and related products such as tires, batteries and accessories. Although the function of a jobber is a very important part of the petroleum distribution system, jobbers have had a low profile to the point that we were shocked to learn that many people in government do not realize either our existence or our role. I am an independent businessman, owning my own trucks and real estate and operating my company in five counties of East Tennessee which are both rural and suburban.

In 1942 my father elected to represent Shell Oil Company as a jobber. Although I have actively considered propositions from competing suppliers, we have never switched brands. Since its founding, we have worked to make this family company grow while seeking two divergent goals. Our first goal has been to grow in sales while still maintaining a correct, independent relationship with our supplier. For example, in the 1950's we borrowed money through our supplier to construct service stations; but in recent years we have looked elsewhere for financing and now owe Shell nothing on such loans. In order to be able to change suppliers on short notice we sought and received in the late 1960's a short one-year supply contract.

Our second goal has been deliberately and meticulously to imitate our supplier in station design, exterior layout, advertising signs and standards of appearance. The purpose is to take advantage of Shell's national image to induce people to buy more from Johnson Oil Comapny. Many of our tourist customers mistakenly assume that they are purchasing at a station belonging to Shell.

That background brings me to the point of my presentation from my viewpoint with many years of experience in this business and particularly for the marketing of gasoline. From firsthand experience I can state unequivocally that the current diverse system of marketing which includes vertically integrated companies produces vigorous competition on the streets and highways of East Tennessee. It existed before the 1973 shortages, and it exists today, with no dealer or jobber of whom I am aware taking the full markup allowed by the FEA.

There is no law or complex set of regulations that in my opinion is going to increase marketing competition. In Morristown, we compete with other jobbers selling Phillips, Texaco, American, Union 76 and Mobil. Gulf Oil and Exxon have agents. Tenneco and Continental are represented by marketing chains, and at least five smaller regional private brands are sold through salaried outlets. Competition is extremely keen, keeping constant pressure on the small businessman to offer a variety of marketing concepts.

Our company owns or leases thirteen full service stations, one truck stop, one tourist restaurant, and three convenience markets. All of these are operated by independent, lessee dealers. In addition we operate ourselves, through an associated company, two combination carwash and self-service gasoline outlets, and we supply gasoline to 39 predominantly rural grocery accounts. We also compete in the sale of heating oil and fuels for commercial accounts.

In our territory it just is not true that major brand outlets are choking out private brand competition. Indeed we have seen just the reverse with an accelerating pace in the last five years. There has been rapid growth in the number and variety of high volume outlets for private brand companies such as Red Ace, Bay and Publix Oil.

Vertical integration has given many large companies a strong incentive to move their refined products under their brand name in our market. Particularly outside the metropolitan areas of Knoxville, Chattanooga, Nashville and Memphis, these majors have worked in Tennessee through local independent businessmen such as ourselves to perform the marketing function for them. It seems obvious to us that divestiture would cause marketing to consolidate into fewer and bigger marketing companies which would displace thousands of dealers and wholesalers with their won new high volume, salaried chains because of their immense leverage to negotiate long-range contracts for refinery production. This revolution may work to the advantage of a few very large super-jobbers and private brand chains which have the money and resources to create their own brand images, but it can only work to the detriment of small jobbers who have relied on their suppliers for major brand identification, image standards, credit terms, promotional advertising, product advertising, station design, engineering services, marketing experience,credit card programs, personnel training and other forms of business support. It would seem obvious that the economies of size would force these new big companies, these "Sears and Roebucks" of gasoline marketing, into existence; and there would probably be no room for the jobber under their umbrella.

I would also predict that after the marketing consolidation does take place and the new giants have been formed, prices at the retail level will then be far easier to manipulate upward than they are today in a field which is still abundantly filled with individualists who aggressively innovate to meet the public's marketing demands. Indeed it would seem that the major reason for not breaking up the vertically integrated companies is price. New layers of management would have to be put on every level of the former petroleum chain; and with the same sureness that night follows day that increase in overhead is going to increase the price of gasoline at the pump. It is an idler's dream to believe that stockholders in the newly dismembered companies would be satisfied with a rate of return on their new investments that is significantly lower than what they now receive.

The trauma of the FEA's immense web of incredibly complex rules is a nightmare that Johnson Oil Company will never forget. Yet, difficult as that has been, the impact on the petroleum business of those regulations would be minor compared to the massive confusion, uncertainty and absolute chaos that would be caused by the implementation of divestiture. Undoubtedly there will be those few giant private-brand chains with the wealth, the inside regulatory knowledge, and the national marketing skills to work this confusion to their financial advantage, but for the small wholesaler I can foresee almost nothing but oblivion.

It would seem to me that the individuals with the most to lose from divestiture would be the independent branded jobbers, such as ourselves, and the independent businessman who leases a branded station. Unlike some forms of business, there is little or no incentive for a new super chain to buy out small companies. They need only to construct a new establishment next door to the current station proclaiming their new image. What would happen to a Shell jobber if that producer suddenly lost its desire to sell Shell branded products in Morristown? Today I would have some choices because other brands do want to market there. The "tomorrow" envisioned by Senate Bill 2387 could very well deny me a competitive choice. The ability of a jobber to associate himself with a vertically integrated company and the ability to change that association to a similar but competing company is the rock bed of his independence and competitiveness. This bill would destroy that rock bed and replace it with guaranteed chaos and a promise of something vaguely better sometime and somehow.

I would like to think that the deliberative processes of the United States Senate would reveal even to those inclined to the publicly popular position of punishment for "big oil" that this country cannot afford the trauma of dismemberment. Divestiture will create new marketing giants, it could injure market competition, it will increase prices, it will curtail investments in the acquisition of energy resources, it will create unparalleled chaos in a fundamental segment of the American economy, and it will be disastrous for thousands of small businessmen. Mr. Chairman, I respectfully urge this committee to reject the concept of divestiture. Thank you.

12

A View From A Small Independent Distributor

TESTIMONY OF ROBERT J. WELSH, JR., PRESIDENT AND OWNER OF WELSH OIL, INC., BEFORE THE SUBCOMMITTEE ON ANTITRUST AND MONOPOLY, COMMITTEE ON THE JUDICIARY, UNITED STATES SENATE — NOVEMBER 12, 1975

Summary: Welsh Oil is an independent jobber operation, associated with Phillips Petroleum Company. Divestiture has already affected its investment—Welsh Oil abandoned plans for building a new truck stop, being uncertain of the future. Divestiture brings many questions. Who is going to get the brand name—the refiner or the marketer? The new marketing company might compete with us under the same brand that we have helped build up over the years. Who will supply us? Who will extend credit and who will support the credit card program?

Robert J. Welsh, Jr. speaks . . .

I am here today representing only myself and my business, Welsh Oil, Inc., Gary, Indiana. The views I have are my own. I suspect, however, that my views are shared by other small, independent distributors.

I am very concerned about recent Senate votes on the question of breaking up the oil industry in this country. I am worried about how a radical restructuring of the industry will affect my ability to serve my customers in Northwestern Indiana. Quite frankly, Senator, I fear the worst.

I am here today because Senate Bill 2387 does not spell out how dismembering the oil industry is going to affect my business. I am not convinced that restructuring the oil industry is just a paper transaction.

Our company was founded by my father in 1925. Over a period of fifty years we have built an association with our supplier and with our customers and with our employees. Welsh Oil is an independent jobber operation. We have 27 service stations that serve customers in Northwestern Indiana. We have 83 employees in our oil business.

There are about 500 other small independent distributors like us in the state of Indiana. Quite frankly, I am worried about what S. 2387 will do to little companies like ours. If you attack our suppliers, you are going to affect our businesses, too.

I mentioned our association with our supplier, The Phillips Petroleum Company. We have a lot of time and effort invested in associating friendly and reliable service with their trademarks and their brand name. It means something to us and to our customers that this beneficial relationship continues.

What's going to happen to that brand name if a bill like this becomes law? Does it go to the refiner or to the marketer? This bill does not offer even a hint of how a question like that would be answered.

And what about the supply of product itself? If all of the controls and regulations are continued as they might be, from whom do I get my supplies of gasoline and fuel oil? Do I get them from a new Phillips marketing company? And if the controls and allocations continue, from whom does the marketing company get its product? A Phillips refining company?

If that's the case, this bill does not really change anything. It will not make the situation any better and it may make things a lot worse. It's going to slow down the whole process of moving oil products to the customer and, Senator, it simply is going to add to my cost and to the price that my customers will have to pay.

There are some other things that scare me too. If you take a major oil company and turn it into four companies, one of them is going to be a marketing company. If it is going to be a marketing company only, it's going to look at ways of being the best and most efficient marketing company that it can. That might just mean it will decide to become a wholesaler at my level. This bill could very well mean that instead of competing with other independent distributors like myself, I would be competing with a big new oil company which might have the brand name that I have helped build up over the years.

You know there are all kinds of assurances by people who do not understand how our industry operates, that this bill is not intended to affect the little jobber or dealer. Personally, I think that *we* are the people who are going to be affected most of all.

When the energy crisis and the oil embargo started, all sorts of controls were put on. They were not designed to hurt the distributor or the dealer, but I invite any of this subcommittee's members to visit my office and see what those regulations have cost us just in paperwork. All of these well intended regulations are what helped bring about those long lines at the gas stations.

Now I have seen some arguments that my supplier has put out against bills like this, and I have read a lot of reports in our trade papers. I do not always agree with everything they say, but they are correct in saying that this bill will affect the oil industry and the energy users of America in profound ways. I agree that there probably could never be a worse time for a bill like this to be considered if we are going to even hope to be less dependent on imports. I agree that the people who are really going to be hurt by this bill are the people who use gasoline and fuel oil because they are going to have to pay the final bill for the government's mistakes.

I agree with all of that, but those are not the reasons which brought me here. My concern is for Welsh Oil, which I believe is a good, small, independent distributor of gasoline, fuel oil, and diesel fuel, with substantial capital invested in real estate to market these products. There are companies like ours all over the country, and I do not believe the subcommittee has given sufficient study to the effect this bill is going to have on us.

Indeed, it is *already* having an effect. In 1972, our company purchased 17 acres for the construction of a new truck stop on the Southeast quadrant of Interstate 65 on Route 2 at the Lowell, Indiana interchange. The energy crisis started shortly thereafter and because of this we were unable to procure additional gas and diesel so our substantial investment has sat idle since date of purchase.

We applied to the Federal Energy Agency and they agreed that a truck stop was needed on I-65 in this area, as the closest truck stop is over 40 miles away.

If we build this truck stop, it is going to provide construction work and means a lot of jobs in that area. It is going to mean the purchase of a lot of building materials and supplies during construction and when completed it will also mean a lot of new jobs in the Lowell area.

The easy way—the way a lot of distributors must feel when they see a bill like this one—is to forget the new truck stop or the remodeling or construction of a new gasoline or truck stop outlet and just get along with what we already have.

Right now, we are in sort of a holding pattern on this project. We don't know whether to go ahead or not. It is a substantial investment for our little firm, and we are not that sure of the future with legislation like this being considered.

I have discussed your bill and other bills to dismantle the oil industry with my employees, customers and other oil distributors in the state of Indiana and believe me when I say there is absolutely *not one* person that I have talked to can understand why Congress would take this action. We all feel that we are in a very competitive industry—and sell our petroleum products in a very competitive market.

As I said, Welsh Oil is just one of about 500 independent distributors in the state of Indiana. We all have a stake in this bill. So do the thousands of independent oil distributors' employees in Indiana who work for us. So do the people who work for the 15,000 independent distributors in all of the states.

I think we all need some answers:

Would we have to depend on new suppliers if this bill became law?—would we have to depend on various suppliers? Would they be willing to extend credit terms to me like my present supplier does? Would they support a credit card program for my customers? Would they continue to advertise the brand name I use, or would I have to give up the brand identity we have developed over a long number of years? These are just some of the questions which need to be explored and answered before you rush into legislation that might be very disastrous to our company.

In this case, I believe you are attacking a system which is trying its best to cope with the energy crisis. I believe that if the government wants to help, it should expand its campaign to conserve energy and to develop new energy supplies, as you personally have suggested on the floor of the Senate yourself, just a few weeks ago, Senator Bayh.

13

International Relations and Divestiture

TESTIMONY OF WILLIAM P. TAVOULAREAS, PRESIDENT, MOBIL OIL CORPORATION, BEFORE THE SUBCOMMITTEE ON ANTITRUST AND MONOPOLY, COMMITTEE ON THE JUDICIARY, UNITED STATES SENATE — JANUARY 21, 1976.

Summary: The oil industry is not concentrated—the largest single domestic crude producer or refiner or gasoline marketer has about 8% of the business. Independents have increased their market share. Profit rates have been about equal to the average return for all U.S. manufacturing. 1974 high profits were due to inventory profits; 1975 profits have declined.

Any refinery or marketer company spun off from an integrated company would have to make a much higher profit than the integrated companies are making today in these sectors. It could only mean higher prices to the consumer.

Because of the large capital requirements, the low rate of return, the inflexibility of the system once constructed, independent companies have not stepped forward to build new pipeline systems.

An integrated company can put money to research work in areas where the need is greatest.

During the embargo, the large oil companies utilized their worldwide logistic to cushion the blow to the U.S. by moving non-Arab oil into the U.S. and Arab oil to countries which were not embargoed.

The proponents of this legislation have not attempted to lay down a blueprint of the consequences. Divestiture could simply become a back door to nationalization of the industry.

William P. Tavoulareas speaks . . .

I am glad to have this opportunity to testify before you today because I think I can shed some light on this issue of oil-company divestiture which is pending before you. Because so many unsupported statements have been made on the subject, I would like to begin with some facts about the oil business.

It is a fact that the world's industrialized countries have based their growth on energy from abundant supplies of low-cost oil.

It is a fact that the United States has remained the world's greatest industrial power by relying on oil and natural gas for three-quarters of its energy supply.

It is a fact that oil, more than any other commodity, helped Western Europe and Japan to emerge from the wreckage of World War II to economic prosperity.

It is a fact that all of this was made possible by an international oil industry which produced, refined, and marketed the oil, using a tremendously complex logistics system which got the right crude oil or product to the right customer at the right time.

It is also a fact that the international oil industry has greatly changed in the last two years. While the oil industry still retains its worldwide transportation, refining and marketing facilities, a significant proportion of the world's crude oil reserves is now directly owned by producing-country governments. These goverments have also clearly demonstrated their ability to raise prices without consulting either oil companies or consumer-country governments, to cut back production when it suits them, and to deny oil to consumer countries whose policies they do not like.

It is also a fact that oil companies recorded far higher profits than usual during the embargo period, giving rise t an "oil industry conspiracy" theory among a public that read about the profits in newspapers as it stood on line for gasoline at much higher prices. I will have more to say about profits later.

All of these developments have had the double effect of depriving the oil industry of much of its control of oil production and at the same time increasing suspicion of oil companies in the industrialized countries. Consumers see only that prices go up, and blame the nearest target—the oil companies. Governments are concerned that they have no weapons to break the control of oil prices by the Organization of Petroleum Exporting Countries—the world's most successful cartel. The result is general frustration and a desire to strike out at someone.

The present moves to dismantle the oil industry in the United States stem at least in part from that frustration. But emotion is a poor guide to action. We need to look at the issues clearly. If we do this, we will be able to talk *to* each other, instead of *past* each other.

Concentration

Let us begin with the issue of concentration. The oil industry has been accused of being a monopoly—whatever that means. I might add that, in the foreign area, no company has more than a 13% share of crude producing, refining, or marketing. Worldwide, my own company has 5 or 6% of the world market.

But rather than look at nationwide concentration ratios, another interesting way of measuring the realities of concentration is to stop and think about the businesses which we come in contact with every day. Most of us have the choice of at least four or five different oil companies when we purchase gasoline; at some intersections there is a different company on every corner. By contrast, how many supermarkets do you have to choose among—two, perhaps three? How many drugstores are there in your community? How many newspapers? Indeed, at the retail level most of the comments I hear criticize the *large number* of service stations available to the public. I must say I don't hear communities complaining because there is insufficient oil company competition for the customers' business.

Indeed, most people who seriously accuse the oil industry of being concentrated seem really to be complaining that the industry is large. In truth, there are many large companies in the industry.

But it is not true that presence of large companies prohibits smaller ones from entering the business and prospering. So-called "independent" companies have, in fact, done very well. Independents have shared in four out of every five winning bids on significant tracts for offshore lease sales. Ten companies which were not in the refining business at all in 1950 had each built refineries with capacity of over 50,000 barrels a day by 1973. Independents increased their share of the U.S. gasoline market from a quarter in 1968 to a third in 1975—a really significant achievement.

In spite of these facts, there are still some critics determined to make the oil industry into a monopoly, even if they have to count up to 20 companies before they can make the charge stick. What the words "20-company monopoly" mean I have no idea. Perhaps it is best to leave it as a mystery.

Profits

If "monopoly" is one myth, high oil-company profits are another myth which has helped give rise to the divestiture movement. Let's look at profits. More accurately, let's look at the oil industry's rate of return on investment.

In the whole period of 1960-74, the oil industry's rate of return on shareholder's equity—the measure of what the shareholder earns on the money he has invested—was 12.3%. This was roughly equal to the average return of all American manufacturing. The rate was significantly *lower* than that for soft drinks, drugs and medicines, soap and cosmetics, office equipment, tobacco, and many other industry groups.

The year 1974 was indeed abnormal, and it is therefore important to understand why oil industry earnings were higher in that year.

Mainly, the profit increases were due to so-called inventory profits. When OPEC raised oil prices, the value of a barrel of oil in storage increased by a factor of four or five times. But these increases didn't give the company any money it could use because it had to replace any barrel of oil taken out of inventory with another barrel of oil at the new price. Yet the *same* amount of oil was in storage and its cost was $12.00 - $15.00 a barrel rather than $3.00 - $4.00 a barrel. The price increase produced substantial reported profits which resulted in *cash* tax payments, while the new high-cost inventories represented a further *cash* outlay. In short, with the same number of barrels of oil in our tanks, there was a cash drain, and we also suffered public criticism for what some have called "obscene" profits.

If you eliminate inventory profits, 1974 profitability was not out of line with the rest of industry even in that year. I should also point out that oil-company profits dropped considerably in 1975. Costs of raw materials and supplies went up. Companies battled against inflation, higher taxes resulting mainly from loss of percentage depletion, and government regulations and controls. The profits of 34 companies analyzed by the First National City Bank dropped by over 30% in the first nine months of 1975 (the latest figures we have) compared with the first nine months of 1974. Between the first nine months of 1973 and the first nine months of 1975, those same 34 petroleum companies increased earnings by only 9%—a gain of roughly half the rate of inflation during the period.

Surely, in view of the evidence, there can be no more concern about "obscene profits" as a motive for dismantling the oil industry.

Size and the Consumer

If the oil industry is not highly concentrated relative to other major industries, and not highly profitable, what is the true concern behind this legislation? Let me suggest that in part it stems from the fact that many oil companies are very large. The industry itself is immense—the 29 oil companies analyzed by the Chase Manhattan Bank have assets of about $193 billion. Yet we often forget that a sizeable part of these assets are overseas and represent investments made to serve markets totally outside the United States. In the case of the 29 companies, those assets amount to $110 billion, or 57% of the total. Obviously, one of the ways to reduce the apparent size of the U.S. oil industry would be to force it to give up its foreign assets; I really wonder if anyone seriously believes that such a result would be in the interest of the United States and its policies. But I would submit that size is in itself no reason for wanting to break them up. In fact, the nation, and particularly the consumer, has a vested interest in *not* dismantling the oil industry as we know it.

Mainly, it's a question of money. Consider some of the figures. Mobil recently spent $350 million for a single producing platform in the North Sea; structures of the same or greater cost will likely be required to exploit areas offshore North America. A new refinery of competitive size could cost up to $800 million. Big companies make these huge investments to hold down unit costs, and if they could not afford to do it, the alternative would be either no supplies or supplies at much higher costs.

Let's take a look at what might happen if integrated companies were broken up into segments. If all the segments are to survive as independent entities, each will have to earn an adequate rate of return. In particular, this would be true of the refining and marketing segments of the business. I leave it to your judgment as to what rate of return is adequate. Pick the figure you think fairest—because I am certain that, whatever figure you pick, it will be much higher than what we make. Mobil lost money on its $2.8 billion of domestic refining, marketing and transportation assets in the first half of 1975, and only began to break even in the latter part of the year. In 1974 the situation was even worse. The point is obvious: Any refining or marketing company which was spun off from an integrated company would have to make a much higher profit than the integrated companies are making today in these sectors of the business. And they could only make that profit by charging the consumer higher prices.

On the face of it, the producing segment of the industry would appear to be the most profitable of all. This attractiveness arises only because past crude oil discoveries are being liquidated at today's prices. In point of fact, new exploration in many of the difficult offshore areas would not be profitable at the controlled crude oil prices which have just been enacted. Thus, even producing companies would have only a limited period of attractive profits if they continued to explore for oil to be sold at today's prices.

Let's look at the subject of pipeline divestiture. Because of the large capital requirements, the low rate of return and the inflexibility of these systems once constructed, independent companies have not stepped forward to build new pipeline systems. Certainly there was no flood of pipeline companies wanting to take the risk of investing in the $6 billion Trans-Alaska line. That pipeline was built solely because companies exploring on the North Slope had to build one to get the crude out. And independent pipeline investors failed to come forward for two reasons—they did not have the money and, if they had, they could have found more lucrative and safer places to put it.

In other areas, too, such as the Gulf of Mexico, major companies have no choice but to built these lines to get crude oil to their refineries, even though rates of return are low, and even though the pipelines have only a limited life because they are useless when once the crude oil they carry is used up.

Look at research. An integrated company with capital available for research can put sums of money to work in areas where the need is greatest—at one time in exploration and producing, another in refining. In Mobil's case, we lead the industry in the development of zeolitic catalysts—which are used to increase the gasoline yield from a barrel of crude oil—which greatly benefited the American motorist. We have put large sums of money into improving seismic detection techniques which considerably improve our chances of finding oil.

If our own government is not convinced of the value of our research, it is ironic that the Russian government has for some time been attempting to purchase American private companies' knowhow for application to its own oil industry. This is a significant example of the inability of centralized government planning of the kind practiced in the Soviet Union to conduct exploration and production as successfully as our own private companies.

Let me talk briefly now about the function of size in our international operations—again as this relates to the U.S. consumer. Look back at the 1973-1974 embargo period. When the Arab nations cut off oil supplies to the United States, the large oil companies utilized their worldwide logistics capability to move more non-Arab oil into the United States and move Arab oil to countries which were not embargoed. As a result of these efforts, the effect of the embargo was much less bad than it might have been. This total industry effort to allocate supplies took place at a time when consumer nations could not agree on a coherent policy, and was given high praise both by the Federal Energy Administration and by the European Economic Commission which exhaustively investigated the companies' behavior during the embargo period. But perhaps the best measure of our success was that we so cushioned the blow to the United States that millions of Americans still say they believe there was no real crisis. If we hadn't been so efficient, we might have been in a better political position today.

In international operations, size is vital, and it is essential if the American consumer is to continue getting supplies. Whatever action the U.S. government takes will not alter the fact that the Royal Dutch Shell group and British Petroleum operate overseas, and you can be sure they stand ready to pick up any pieces which may fall to them as the result of legislation here.

It is therefore especially surprising to note that there is now a theory making the rounds that the United States will actually be better off if it breaks up its major oil companies, because OPEC is able to maintain its grip on crude prices only because it has a subservient network of oil companies to market its oil.

Let me say bluntly that this kind of thinking is exactly 180 degrees off target. There is no denying the fact that OPEC controls the crude supply. Yet the international companies, with their refineries and markets around the world, afford the consuming countries a diverse source of supply for crude oil and products. No single country could develop such access on a basis unfettered by political considerations. Many foreign countries—such as Japan and those in Western Europe—recognize the value which the international companies represent, and work with these companies instead of against them. Britain, for example, utilized the expertise of the international companies to find oil in the North Sea, and that nation will be self-sufficient in oil in the 1980's. To sweep all of that away and replace it by a host of dismembered companies will not produce a better value for the United States—quite the reverse is true. The basic fact we have to live with is that—until we can substantially reduce our dependence on foreign oil—we will have to pay the OPEC price for imports. No policy of splitting up the oil companies is going to help the consumer.

Indeed, from all the considerations I have mentioned, the American consumer would have to foot an immense bill for divestiture, in return for absolutely nothing at all or—more accurately—in return for the privilege of paying more for far worse products.

In preparing to testify before you today I reviewed some of the testimony which this Committee has received, and was struck by the fact that the proponents of this legislation have not attempted to lay down a blueprint of the consequences of this legislation. Such a blueprint is particularly essential, for we are dealing with a situation which is unusual in American practice. When divestiture is ordered by decree of the court following a judicial conclusion that the antitrust laws have been violated, the law breaker has little basis to complain that his business is being disrupted. Yet here there has been no finding that the law has been broken, and indeed the long years of activity by the Justice Department and the Federal Trade Commission strongly suggest that quite the opposite is true; instead, dismemberment can apparently be accomplished *only* by affirmative legislation. In those circumstances, it would seem to me especially incumbent on the proponents of the bill to state explicitly what they hope to accomplish.

Do they hope that the price of foreign oil will decline when a host of smaller American companies compete with the foreign giants for scarce supplies of foreign oil?

Do they, on the other hand, hope that the price of gasoline will decline at the service station when a host of American companies have to earn a reasonable—let us say even a meager—rate of return on the tremendous assets involved in marketing?

Or do they hope that the price of oil will decline when smaller American refining companies have to build 50,000 barrel-a-day refineries rather than 150,000 barrel-a-day refineries because they do not have the capital resources to make the larger investments?

Or possibly do they hope that supplies of oil will increase and its costs will come down when independent pipeline companies provide the pipeline transportation *only* after the reserves to support the line are fully developed and when the financial institutions are convinced that the accumulations are large enough to make the pipeline viable over its useful life?

I cannot give positive answers to any of these questions. I do not know whether the other witnesses you will hear can do better.

Alternatively, divestiture would simply become a back door to nationalization of much of the industry. As we all know, government control of industry means red tape, it means bureaucracy, and above all it means higher prices—as a recent comparison of private international companies and government oil companies has amply demonstrated.

This last point is one that worries me a great deal. I do not think that the honest—though misguided—proponents of divestiture see it as a roundabout way to government control of the oil industry. But there are those who would see the failure of American industry to find and produce needed supplies as a welcome assist to their advocacy of enforced "conservation" and increased government intervention in oil exploration and production. The fact that an industry dominated by big government would be the result is the ultimate irony for all those now concerned that oil companies are too big.

To sum up: Proponents of divestiture have, to my knowledge, produced *no* evidence that it would benefit the country. On the contrary—they have at best suggested a method of weakening the oil industry when a *strong* oil industry is desperately needed—to raise the billions in capital requirements that will be needed to improve our energy self-sufficiency, and to reduce instead of increase our vulnerability in an emergency, whether it is another embargo or a war in which the United States is involved.

By long years of service, American oil companies have demonstrated that—when governments do not suffocate their enterprise and efficiency—they can supply customers around the world with excellent products at reasonable cost. I hope we can continue to do the job.

14

Divestiture and OPEC, A Response to Anthony Sampson

STATEMENT OF STANDARD OIL COMPANY OF CALIFORNIA CONCERNING THE TESTIMONY OF ANTHONY SAMPSON BEFORE THE SUBCOMMITTEE ON ANTITURST AND MONOPOLY, COMMITTEE ON THE JUDICIARY, UNITED STATES SENATE — FEBRUARY 3, 1976.

Summary: Anthony Sampson states that the oil companies did not encourage the 1973 oil price rises, but that the central issue is that there is no near-term substitute for OPEC oil. The oil companies do not prorate production for OPEC. OPEC's tool is not allocation of production but coordination of price and taxing policies. The oil companies are providing both some stability and diversity in supply as well as technology, business management and capital. They also make it possible for consuming countries to receive the benefit of OPEC price discrepancies which occur from time to time as a result of changing freight rates and product mix. Mr. Sampson describes U. S. foreign policy as assisting OPEC countries in their industrialization plans. The U. S. considers the oil companies as instrumentalities for the fulfillment of this objective.

It is interesting that a Subcommittee of the United States Senate should look to a foreign journalist who by his own admission is neither an economist nor an oil specialist to testify on the economics of oil. Indeed, his testimony exhibits a greater flair for fiction. Mr. Anthony Sampson's claim to any knowledge in regard to the issues on which he testified on February 3, 1976, before the Subcommittee on Antitrust and Monopoly of the

Senate's Committee on the Judiciary apparently stems from various statements and comments which are attributed by him to certain Middle East oil personalities, hardly disinterested observers of the international oil scene. One must wonder whether Mr. Sampson has not been taken in by OPEC propaganda. What could be more natural than for OPEC representatives to point the finger of blame at some other conspicuous target.

We set forth below the *facts* in regard to the issues raised. In doing so, we concentrate on the basic thesis and ignore the peripheral areas raised by the testimony, including the subject of multinational corporations generally, bribes, the farsightedness (or lack thereof) of the Boards of Directors of oil companies, the needs, if any, for public directors on such boards, and the like. Although these may be interesting and important areas for debate, they do not constitute the main issue.

In its simplest form, the Sampson argument is as follows:

1. *Problem* — The *number* and *size* of oil companies having crude oil dealings with the major producing countries of Saudi Arabia, Iran and Kuwait allegedly led OPEC to organize in the first place and facilitate its continuance as an effective cartel.

2. *Proposed Cure* — Increase the number (and presumably the mix of sizes) of oil companies buying crude oil from those producing countries.

3. *Claimed Results* — Greater opportunities for independent oil companies, increased competition in buying crude oil from the major producing countries, less vulnerability for consuming nations to the OPEC cartel, and greater flexibility to the U.S. in its Middle East foreign policies.

This basic thesis is brought into sharper focus because even Mr. Sampson admits what so many others (even a few Presidential aspirants) have finally conceded—that the companies did *not* sponsor or encourage the 1973 oil price rises in order to increase their profits and that *no conspiracy* existed or exists between the companies and OPEC. He also concedes that however one examines the situation, the U.S. will be increasingly dependent upon oil from the Middle East in the near and intermediate term, a fundamental and inescapable fact of life on which there is at last almost universal agreement. We shall have more to say about this later. Additionally, we strongly concur that it would be unacceptable over the long term for consuming nations to allow themselves to be entirely dependent upon OPEC oil or its unilateral pricing policies and that some way must be found to accord consumers and third world nations a strong voice with the producers in pricing matters.

This is one of the basic objectives of the International Energy Agency and of the Paris meetings among representatives of those groups. Finally, it is evident, as Mr. Sampson agrees, that nationalization would not provide a more efficient oil business and that the oil companies have been unfairly blamed for the faults of governments and consumers.

Dependence on OPEC

Unfortunately, the value of these positive conclusions is lost when one attempts to reconcile them with the "multiplicity of companies" idea. The central issue of world oil today is *not* the number and/or size of companies having crude dealings with major producing countries nor is it any other simplistic structural design; *it is the simple fact that there is no near term or intermediate term substitute for OPEC oil.* The plain fact is that today the great bulk of non-communist world's oil reserves happen to be in the OPEC countries, primarily in the large oilfields of the major oil producing nations of the Persian Gulf. No amount of realigning, restructuring or tinkering is going to change that fact. The reserves on which the free world must rely will still be where they are whether one company or 100 companies are set up to acquire crude oil from the countries owning them.

It is difficult to discern why more buyers of crude from OPEC countries would be advantageous to consuming nations. What is it about 100 buyers that would change the present situation to the benefit of consumers? Can 100 new oil companies suddenly spring forth out of the earth with financing, ships, refineries, terminals, pipelines, bulk plants and service stations to move oil from these countries across the oceans of the world and turn them into essential products to fuel the industrial economies of the free societies? The notion, although somewhat obscure, seems to be that OPEC will find it easier to hold "hostage" a few large buyers. Because oil is important to these companies, it may force them to act as "agent" of the cartel in carrying the "prorationing" of production which OPEC has never been able to accomplish itself. Of course, the fact is that the oil is important to the consuming countries. These countries could not become indifferent to the availability of this oil if the major companies somehow vanished. Access to the oil is essential to economic survival and given the enormous increases in its price imposed by OPEC it is essential that it reach consumers in the most efficient and least expensive way.

How OPEC Operates

The "production prorationing agent" hypothesis suggests a lack of understanding as to how OPEC operates its cartel. OPEC does *not* attempt to prorate production quantitatively either directly or through the companies. It has had no present need for production prorationing because it is able thus far to achieve its objectives quite satisfactorily from its standpoint through its coordination of unilaterally imposed prices and taxes.

Since the end of the 1973-1974 Arab production cutbacks, the OPEC governments have not collectively set production levels. There has been no "orchestration" of production rates by oil companies or anyone else. Production has just "dropped"—not been "cut"—and dropped by different amounts in different countries, simply because consumer *demand* has been weak as a consequence of the worldwide recession, conservation efforts and the extraordinarily high oil prices. The practice of the major producing companies both now and throughout most of their history has been to provide capacity, and spare capacity too, to meet all the demands and requirements of a highly competitive market. Production rates are not established in advance—they are only known after the fact. If another tanker had called for crude oil at any of the major Persian Gulf loading ports on December 31, 1975, it would have been loaded—and this would have been true regardless of the ownership of the ship or the identity of the customer.

No, OPEC's tool today is not allocation of production; it is coordination of price and taxing policies. The fact that there may be a plentiful supply of crude oil does not put pressure on the government to reduce its prices and taxes as long as the government is content with the volume sold and its resulting total revenues. Thus, as long as the volume effects are considered to be tolerable by individual OPEC states, OPEC price/tax coordination is not susceptible to competitive erosion from surplus supply in the manner that business profits are.

Coordination of pricing and taxing policy has proven simpler for OPEC and less fraught with politics and emotion than coordination of production policy. This has been true notwithstanding the fact that OPEC's coordination of this policy has been less than perfect and to some extent unpredictable in its results. During the recent period, for example, changing freight and quality factors effectively caused situations where some OPEC countries were receiving relatively less revenue per barrel and other countries were receiving relatively more, with the result that some countries (e.g. Iraq, Indonesia) approached full capacity production while others (e.g. Saudi Arabia, Libya) drifted down. This coordination mechanism works as long as there are enough "drifters down" willing to accept less production but more money.

To believe that the companies are in some way orchestrating production or regulating the market is to be gulled by OPEC propaganda. Few companies have significant operations in more than two or three OPEC countries—and no company operates in all. Orchestration of production in all 13 OPEC countries, controlling over 80% of free world reserves, would thus be as miraculous as stopping the earth on its axis. The market in fact is dynamic and competitive, as demonstrated by the uneven production swings by the different producers and the fact that products are available in Europe at marginal cost, consisting of required payments to producing governments, spot freight and incremental refining and distribution costs.

The OPEC countries, acting in concert, are able through their price/tax coordination policy to control prices, and they are able to do so *alone.* They do not need the oil companies or anyone else to serve as their agents in this regard. Taxes and royalties have risen to the point that Mideast OPEC government "take" is about 95% of total revenues realized. The major companies are not favoring OPEC over the consuming countries. They are providing both some stability and diversity in supply and market as well as supplying technology, business management, and capital. They remain the indispensable bridge between oil in the ground, primarily in the Middle East, and the far-flung markets of the world. Additionally, they make it possible for consuming countries to receive the benefit of OPEC price discrepancies which occur from time to time as a result of changing freight rates and product mix. An example of the latter has been the recent changing relative values of light and heavy crudes as fuel oil demand changed. The market draws crude preferentially from areas providing the most attractive economic package.

The consuming countries are dependent on OPEC oil today and for the immediate future. That is the basic problem and none other. Once one understands and accepts that, all of the current accusations, charges and countercharges involving the oil companies fall. Loyalty of the companies? Of course the American oil companies are loyal to the United States, as has been proven time and again—but they *also* are compelled to obey the laws of the foreign countries in which they operate, just as our own government expects foreign firms operating in the United States to comply with American laws. Does anyone really suggest that American firms should obey *only* American laws in their worldwide operations? What of U.S. foreign policy? That policy is based on empirical and strategic considerations, *one* of which is the growing need of the United States for imported oil. Unfortunately, the oil exists *there*, not here; and our nation is doing too little to develop more here.

Would the Subcommittee really want to see Aramco replaced in Saudi Arabia by a consortium of Japanese and European oil companies? Or perhaps British ones?

In the end, although Mr. Sampson attempts to disavow it, his thesis is a frontal attack on the foreign policy of the United States. The United States has made agreements to assist Saudi Arabia and Iran in their major plans for industrialization. The large oil companies are considered by both the United States and the producing countries as major national assets and instrumentalities for the fulfillment of these agreements. Now, Mr. Sampson proposes that they should be forbidden such a role. One can only guess at what ultimate purpose he has in mind.

We do not need further confusion in the energy debate—it simply adds to the sterile name calling and fault finding which have paralyzed the adoption of any sensible national plan of action. We *do* need understanding, steadiness of mind, and courage to make the essential decision—the decision to develop our own supplies at home of conventional and unconventional energy sources. Only such a large and sustained addition to our energy supplies will remove the United States from the grip of OPEC and reverse the rising and alarming growth in our dependence on outside sources.

15

Developing Alternate Sources of Energy

TESTIMONY BY W. T. SLICK, JR., SENIOR VICE PRESIDENT, EXXON COMPANY, U.S.A., BEFORE THE SUBCOMMITTEE ON MONOPOLIES AND COMMERCIAL LAW, HOUSE JUDICIARY COMMITTEE — SEPTEMBER 11, 1975

Summary: Concentration of oil company activity in nonpetroleum fuel production is low, and oil company activity in these fuels has increased competition. Exxon's efforts at diversification are motivated by a desire to increase efficiency and the domestic supply of energy, and by bringing to bear its ability to raise capital and investments in related technology and management skills. Oil company diversification into other fuels has resulted in increased output, increased productivity and efficiency, increased research and development, and increased competition.

W. T. Slick, Jr. speaks . . .

I. Introduction

Energy supply problems have led to an immense amount of energy-related legislative proposals over the past two years. Some of these legislative initiatives have been concerned with ways to increase domestic energy supplies and to conserve energy. Many other legislative proposals have sought to tax the energy industry, to regulate it further, and to restructure it. One group of legislative proposals seeks to ban any company with oil or gas production and refining operations from having an interest in coal, coal synthetics, uranium, oil shale, solar and geothermal energy. Exxon Corporation is strongly opposed to such legislation.

Exxon believes that, upon the basis of the facts, Congress will find that no curtailment of oil company diversification is required or desirable. The facts will show clearly that firms within industries involving oil, gas, coal, uranium, and other energy forms are competitive. Furthermore, the facts will show that oil company diversification into alternate energy fields has increased both efficiency and competition within these industries and will continue to do so. The facts also will demonstrate that such diversification by oil companies was a completely logical extension of technological and management skills. In our own case, Exxon recognized that the development of alternate energy resources was an extremely complex, risky, and expensive, yet vital, need for the U.S., and we believed that we had the expertise and resources necessary to help satisfy this need and the willingness to take the risks to invest. Finally, we believe that the facts lead to the conclusion that precluding oil company participation in alternate energy sources would slow down research and development, diminish the number of competitors, and reduce future production of existing and new energy sources which are vitally needed to sustain a prosperous and secure U.S. economy.

II. The Facts

It is necessary to understand the energy supply/demand situation in order to put into proper perspective the role of oil company diversification in supplying the nation's energy needs. In brief, Exxon projects that total U.S. energy demand will increase from an estimated 36.4 million barrels per day of oil equivalent in 1975 to about 56 million barrels per day in 1990. This is an average annual growth rate of about 3.0% a year, significantly below the average growth of 4.0% a year which occurred during the 13 years prior to the Arab oil embargo of 1973-74, and reflects changing consumption patterns and more efficient use of energy brought about by conservation and higher energy cost. Despite this lower growth rate in demand, the U.S. must develop all domestic sources of energy to the fullest in order to limit the growth in reliance on foreign oil imports. There has always been an inseparable link between economic growth in the U.S. and abundant supplies of secure energy. Thus, unless the U.S. achieves an adequate level of domestic sufficiency in energy, the nation is running a very real risk of forcing severe reductions in economic activity—with a corresponding loss of jobs, a decline in individual income, and a reduction in Gross National Product.

Increased conventional oil and gas supplies from future discoveries, for many years, are expected to be barely sufficient to offset the decline in U.S. production from existing oil and gas reserves. Domestic supplies of nuclear energy, of coal, and of synthetic fuels from shale oil and coal are projected to become increasingly important energy sources in the future. To meet these projected demands, Exxon estimates that in 1990 domestic coal output will have to approach 1.1 billion tons a year, nearly twice the 1975 production. About 9% of the coal production in 1990 would be used to produce synthetic oil and gas. Synthetic oil from coal combined with that eventually expected from shale oil will amount to about 3 percent of total oil demand by 1990. Synthetic gas from coal will account for about 5 percent of total gas supply in that year. Similarly, nuclear capacity is expected to increase almost eightfold between 1975 and 1990 for an ambitious growth rate of 14.4 percent/year.

Slower-than-forecast growth in coal, nuclear, or gas would translate directly into additional oil imports, which Exxon already estimates will climb to over 9 million barrels per day in 1978, over 12 million barrels per day in 1990, and will account for about 50 percent of total oil demand throughout the 1980's. Clearly, any legislation which would decrease output and efficiency in these sectors will have significant adverse effects on future national energy supplies and the prosperity of the nation's economy.

The enormity of the task of domestic energy development facing the nation is underscored by a review of the implied needs in terms of physical facilities and capital requirements. From 1975 to 1990 cumulatively, Exxon has estimated that the U.S. will need over 170 large coal mines (5 million tons per year each); 1,900 unit trains (100 cars each); 10 oil shale synthetic plants (50 thousand barrels per day each); 10 coal gasification plants (250 million cubic feet per day each); 4 coal liquefaction plants (50 thousand barrels per day each); and 40 uranium mines and mills (2 million pounds per year each); all in addition to the more than 425,000 oil and gas wells and 48 refineries (150 thousand barrels per day each).

The capital requirements associated with these facilities alone amount to $23 billion per year in constant 1975 dollars, and these figures do not even include expenditures for chemicals and marketing facilities. This $23 billion figure compares with an average of approximately $7 billion per year in industry capital expenditures for the period 1963-1973. When marketing and chemicals facilities plus electric utility requirements are added, the figures are even more staggering. For example, the First National City Bank of New York estimated capital requirements of $65 billion per year in constant 1970 dollars and the Secretary of the Treasury estimated requirements of $77 billion per year in constant 1974 dollars.

It should be clear from these figures that the energy market is so large that no one firm or group of firms could possibly monopolize or control the entire energy market or any individual fuel in it. Moreover, it should be clear that, even with significant demand conservation efforts, it will be necessary to utilize *all* available domestic exploration and production expertise, and unprecedented levels of capital will be required from many sources.

Concentration Ratios

On January 21, 1975, Exxon submitted a statement[1] to the Senate Judiciary Subcommittee on Antitrust and Monopoly demonstrating the high level of competition in the petroleum industry. That submission provided a detailed economic analysis of the structure and performance of the petroleum industry, including concentration ratios, profitability, price behavior, degree of vertical integration, ease of entry, technological progressiveness, the independence of action of the larger firms, and the economic rationale for such characteristics of the petroleum industry as exchange agreements and joint bidding. We believe the facts represent irrefutable evidence of the competitiveness and efficiency of firms in the oil and gas industry. Simply put, there is no monopoly power in oil and gas to be "transferred" to coal, nuclear, or other energy sources as some have alleged. To the contrary, firms in the petroleum business are a positive force in improving the economic performance of the coal, nuclear, and synthetics industries, and can continue to be unless prevented by anticompetitive legislation.

Our estimates of concentration ratios in each energy source and in total energy are given in the following table:[2]

Concentration Ratios in Production (%), 1974

	Top 1	Top 4	Top 8
Oil (net liquid hydrocarbons)	9	26	42
Gas	9	25	38
Coal	11	27	37
Uranium (mill production)	17	58	85
Total Energy (BTU Basis)	8	21	34
All U.S. Manufacturing Average	—	39	60

[1]See chapter 2.
[2]Source: As shown in Exhibit 1.

The top four companies accounted for only 21% of total energy production on a BTU basis in 1974 and the top eight had only 34%. Concentration ratios for total energy and for oil, gas and coal are low both in absolute terms and in comparison to the U.S. manufacturing average. Total energy concentration ratios are lower than the individual fuel ratios because, as will be shown, the top firms in one energy area are not the same top firms in other energy sources. Thus, the leading energy producers have quite different interests.

This is shown more clearly in Exhibits 1 and 2. Exhibit 1 lists the top 16 companies by name for 1974 in crude oil and natural gas, in coal, and in uranium production. Since oil and gas are frequently found together, there is a natural correlation between the top producers of oil and the top natural gas producers and so these two energy sources were combined. But no such correlation exists between the oil and gas column and the columns for coal and uranium. A total of 48 companies could potentially appear if no overlap existed between the top 16 oil and gas, coal, and uranium producers. If all the top oil and gas producers were also the top companies in coal and uranium, only 16 companies would appear on the list. In actuality, a total of 42 companies appear, which shows how little overlap exists among the top energy producers. This is a large number of actual or potential competitors, yet it still excludes companies such as Amerada Hess, Pennzoil, Superior Oil, and many hundreds of others who compete effectively in oil and in other energy fields.

To underscore this point, Exhibit 2 lists the top 16 energy companies for 1974, in order of total BTU's produced, followed by their ranking in oil and gas, coal, and uranium production. A dash indicates the company has no production at all of that individual fuel; an asterisk indicates the company has some production, but is not in the top 16 producers. Since oil and gas are 75% of the energy consumed in the U.S. and 72% of domestic energy production, most of the top energy producers are top oil and gas producers. Peabody and Kerr-McGee are the exceptions. But most of the top energy producers are *not* top coal and uranium producers, as is clear from the number of dashes and asterisks in these columns. Only one company (Continental Oil) is among the top 16 in all three individual energy sources, but even then it is only fourteenth in oil and gas production. Only three other companies (Exxon, Gulf and Getty/Skelly) appear both in oil and gas and as one of the top 16 producers in one other fuel.

While Exhibits 1 and 2 describe our estimated ranking of actual energy producers in 1974, they do not show probable new entrants into coal and uranium in the future. While Texaco, Shell, Arco, and Sun currently have no coal production, these companies have acquired coal reserves and can be considered likely new entrants. Indeed, Shell, Arco, and Sun have already contracted to sell coal from their reserves when the mines are completed. Similarly, in uranium, Gulf, Mobil, and Phillips appear to be undertaking significant exploratory programs; they and many others are potential competitors to the firms currently listed in Exhibit 1 as uranium producers.

As these data show, the charge that oil companies' diversification into coal and uranium is anticompetitive because of its effect on increasing concentration in the total energy market is simply not true. Furthermore, the Federal Trade Commission staff's own report on concentration in the energy industry, published January 1974, concludes that such entry has not had any significant effects on energy concentration.[3] Using 1970 data, the FTC found that the concentration ratio for the top four energy producers was 23.4%. If the coal and uranium production of the petroleum companies in the top four energy producers were excluded, the concentration ratio was 23.1%. The difference of 3/10 of one percentage point indicates the insignificant effect of oil company participation in coal or uranium on energy concentration. The difference in the top eight ratios was found to be 8/10 of one percentage point (37.8% versus 37.0%). The FTC staff report concludes (page 236): "The impact on production concentration due to coal and uranium acquisitions by petroleum companies appears to have been very small, at least up to 1970." Examination of total energy production data for 1974 indicates that nothing has changed since 1970 to alter this finding.

Extent of Oil Company Diversification

Coal. What matters most in determining the degree of competition within an industry such as the coal industry is the number, relative size of firms, and ease of entry, not whether coal companies are owned by oil companies. Some 4,000 coal companies were engaged in bituminous coal production totaling 592 million tons in 1973. Many of these companies were divisions or affiliates of other companies; on a common ownership basis, the 4,000 companies can be consolidated into 529 groups of completely independent producers, still a large number. Thus, even if all leading coal companies were owned by oil companies, the coal industry would still be competitive given its low degree of concentration. (See table on page 34.)

Petroleum companies and affiliates operating in coal in 1974 accounted

---

[3] "Concentration Levels & Trends in the Energy Sector of the U.S. Economy," Staff Report to the Federal Trade Commission, January 1974.

for less than 20% of total coal production. Low as this percentage is, this type of statement still implies a greater degree of participation by major oil companies than has occurred. A more detailed look at the seven oil companies with current coal operations and their 1974 output is shown in Exhibit 3. Four of these seven companies (Sohio, Ashland, Belco and Occidental) produce less than half a percent of domestic oil output each. While Gulf and Exxon are substantial oil and gas producers, they have only 1.3% and 0.4% of coal output respectively. The seven companies listed in Exhibit 3 produce about 19% of the oil in the U.S. (Note: The 1974 coal production data is presently available for only the larger companies. The seven companies shown in Exhibit 3 probably include over 95% of oil company coal production, however.)

In coal reserves, the top four companies in the industry had 17% of total reserves in 1974 and the top eight held about 25%. The coal reserve concentration ratios in this submission use a total reserve base of 250.0 billion tons for the lower 48 states.[4] Of this 250 billion tons, about 150 billion tons or 60% is open acreage; that is, acreage not under coal leases to any company. While the Federal government at present has a moratorium on leasing of Federal coal lands which comprise about 90 billion tons of total reserves (of which only about 17% are currently leased), a large amount of open acreage is potentially available to anyone who wishes to risk his investment in the coal industry. The Federal government alone holds more coal than the top 12 companies in the industry combined.

Petroleum companies had about 16% of total coal reserves, as shown in Exhibit 4. This broad brush statement again overemphasizes the extent of overlap between oil and coal companies. Several of the oil companies listed in Exhibit 4 have only minor shares of U.S. oil production. While an oil company (Continental) is the largest coal reserve owner, it has only 5.4% of total coal reserves. The next three largest reserve owners are all non-oil

---

[4]The 250 billion tons of reserves is derived from a Bureau of Mines publication titled "Demonstrated Coal Reserve Base of the United States on January 1, 1974" published June 1974. Exxon in previous analyses has used a total reserve base of 188 billion tons derived from U.S. Geological Survey bulletins for beds thicker than 42 inches at depths less than 1000 feet and assuming 50% recovery of coal in place. The Bureau of Mines data are more recent, more detailed, and probably more accurate since they identify the coal resource by state, by type of coal, and by classification as underground or surface mineable coal (whereas the U.S.G.S. estimate was a single sum). The recent Bureau of Mines report has a total of 434 billion tons of demonstrated coal reserves. If underground recovery is taken as 50% and surface mine recovery as 80%, U.S. reserves of economically recoverable coal meeting the Bureau of Mines specifications would be 250 billion tons in the lower 48 states. Coal deposits in the U.S. including thinner seams and deeper depths are many times these amounts.

companies and Exxon is fifth largest with about 3.0% of reserves. Most of the oil companies listed in Exhibit 4 have less than 1% of total coal reserves each. Since many oil companies have only recently acquired coal reserves, and since coal mines take up to 3-5 years to develop (not counting delays in obtaining permits and licenses), more companies hold reserves than produce coal at the moment. Three of the six oil companies holding reserves but having no 1974 production have already made contracts to sell coal to utilities. These are Sun, Shell, and Atlantic Richfield. While oil companies held 16% of coal reserves in 1974, they produced about 18% of coal production.

Given these data, it is clear that oil companies cannot hold back development of other fuels and have not done so. First, there is a large amount of unleased coal which is potentially available for development by any existing company or new entrant into the coal industry. Second, most coal is sold on long-term contracts which were willingly entered into with customers that desire long-term fixed commitments to specific production levels. If oil companies operating in coal were to combine in an attempt to raise coal prices and restrict coal production, not only would they violate the law and subject themselves to damage suits for breach of contract, but competitors in the remaining 80% of the coal market would simply gain more business as spot buyers switched suppliers and new contracts were awarded to other competitors. Competition has regulated and will continue to regulate the market provided that anticompetitive legislation is avoided.

Another reason why oil company participation in other energy sources cannot slow the development of other fuels is that oil and other fuels are often not substitutable, especially in the short run. For example, automobiles by and large must run on gasoline. In residential and commercial markets, opportunities for fuel substitutions are quite limited, particularly once the initial equipment is installed. Nuclear power plants cannot switch to coal or oil. Thus, when oil companies diversify into coal or uranium, interfuel competition is not reduced since interfuel substitutability often does not exist. This is not to say that major long-term shifts in fuel use have not occurred and will not occur in the future; witness the substitution of diesel fuel for coal in locomotives. The point is, however, that interfuel substitutability is limited in many markets in the relevant time frame and therefore oil company diversification does not reduce interfuel competition. To the contrary, the entry of new participants in coal, regardless of whether they are oil companies, increases competition in that fuel and increases interfuel competition in the long run.

Uranium. ERDA reported that domestic U_3O_8 production totaled 23.1 million pounds in 1974. About 17 companies were involved in uranium mill production, and many more are currently exploring for uranium. Our estimates show that in 1974 five oil companies represented 32% of uranium mill production as seen in Exhibit 5. Excluding Kerr-McGee (which produces less than half a percent of U.S. oil), oil company participation in uranium drops from 32% to about 17%. If Pioneer Natural Gas Company is also excluded (Pioneer has very minor amounts of oil production), oil company participation falls to less than 15%. The five companies listed in Exhibit 5 produce less than 17% of U.S. oil production. Exxon ranks fifth in uranium milling with 9.1% of 1974 industry production, all of which came from one mine and mill in Wyoming.

We estimate that the top four companies in uranium production listed in Exhibit 1 accounted for about 58% of mill output of U_3O_8 concentrate in 1974 (see table on page 220). Three of these four companies are not in the oil industry and the fourth, Kerr-McGee, is not a large oil producer. In 1974, ERDA reported that 83 companies conducted uranium exploration drilling; only 17 of these 83 companies are presently engaged in uranium mining and milling. Twenty of these 83 companies were oil companies, of which only 5 are presently engaged in mining and milling. Included in the 20 are independent producers such as Canso Oil & Gas, Dalco Oil, Burmah Oil & Gas, and Louisiana Land & Exploration. Thus, many new investors are seeking to enter uranium mining and milling. How many will be successful in their exploration effort cannot be forecast. But it is evident that uranium is a highly competitive industry and concentration in this young and growing industry will undoubtedly decrease as new explorers initiate production and as new sales become more frequent in response to greater installation of nuclear capacity.

According to our estimates of industry uranium reserves, the top four companies accounted for about 57% of total reserves in 1974. Oil companies accounted for about 53% of uranium reserves (33% if Kerr-McGee is excluded). Exxon holds about 6% of industry reserves. The disparity between oil companies' uranium reserves and their production results from the fact that most oil companies entered the uranium business relatively recently and have not yet had time to initiate production. The disparity will diminish in time as mines and mills are opened to meet growth in uranium demand.

Exxon expects that market demand for uranium in 1985 will be almost three and one-half times larger than the 1975 level. An ERDA[5] task force

---

[5]*Nuclear Fuel Cycle, a Report by the Fuel Cycle Task Force*, ERDA, March 1975.

review early this year concluded that the future use of nuclear energy could be limited by the availability of adequate uranium resources. Because it typically takes 8 to 10 years to discover and then develop a uranium mine and associated mill, U.S. self-sufficiency in uranium through 1990 will depend on the success of exploration undertaken in the 1970's. From 1966 to 1974, a period of fairly continuous, intensive exploration, an average of 60 million pounds per year was added. However, a major new discovery has not been made since 1969. Exxon estimates that in order for the domestic mining industry to supply fuel for the reactors forecast to be built by 1990 and maintain 8 years forward reserves, uranium reserve additions would have to average significantly higher than 60 million pounds per year between 1975 and 1990. The meaning is clear. For the nation to establish a high degree of uranium self-sufficiency, it will be necessary to utilize *all* available domestic uranium exploration expertise. Capital will be required from many sources and more companies, not fewer, should be encouraged to explore. There is much opportunity for new entrants and each year new explorers are joining in the search for uranium resources. Even the utilities (TVA, Commonwealth Edison, Southern California Edison, Texas Utilities) have become active participants in recent years. Oil companies are logical participants in this trend.

Other energy sources. While oil company diversification into coal and uranium has drawn the most attention, some legislative proposals would also ban oil firms from participating in oil shale, synthetic oil and gas, and solar energy. Today, this ban would mainly prohibit oil company participation in R&D projects and pilot plants in these areas since no significant economically viable industries have yet been demonstrated to be feasible. Significant production from these energy sources is still many years in the future and depends on additional successful R&D and full-scale commercialization. For example, 33 years have passed since the demonstration that controlled nuclear fission was possible, yet nuclear power accounts for only 2% of the total U.S. energy supply currently. Potential energy sources such as solar power and the breeder reactor are not expected to be commercially available on a significant scale until the 1990's.

While there are no meaningful concentration ratios in these other energy sources, nonetheless, some data exist on the extent of oil company participation in these new energy forms and in energy R&D. Exhibit 6 shows energy R&D expenditures by various industries for 1973. Petroleum companies spent $325 million on energy research in 1973, while all companies spent a total of $875 million on energy research that year. Thus, oil companies accounted for about 37% of all energy research. Firms in many other industries such as electrical equipment, chemicals, metals, and motor vehicles (in "other" category), conducted significant amounts of R&D in energy. About 65% of the $875 million was spent on research in nuclear, oil shale, coal, solar, geothermal; the remaining 35% was oil and gas related research.[6] Of the $875 million, company funds accounted for 68% ($593 million) and government funds for 32% ($282 million). Almost 93% of the government funds were for nuclear R&D, and accounted for 67% of the total funds spent in this energy field. Research in fossil fuels (oil, gas, shale, coal) was almost 98% private funded. Obviously, while oil companies perform a substantial amount of R&D in energy, many other firms in other industries also participate.

Research conducted by Exxon in the U.S. accounted for about 18% of petroleum industry R&D expenditures in 1973. Thus, Exxon's share of R&D expenditures exceeded its share of oil and gas production by a substantial amount. It is to be expected that large companies conduct proportionately more R&D activity than smaller companies since R&D is risky, requires the development of research staffs, and financial results often appear only after many years of substantial investment.

Oil Shale. Oil shale from deposits in Colorado, Utah, and Wyoming represents a potential source of supplemental liquid (and gaseous) fuels which in size is many times the proved domestic reserves of crude petroleum. Because the final product from retorting shale rock and processing the recovered liquids can be a high grade crude oil completely interchangeable with natural crude as feed to petroleum refineries, oil companies have been interested in this resource for 50 years or more.

Since World War II, interest in oil shale has heightened and research and development efforts, financed for the most part by oil companies, have brought technology to a point that demonstration plants could be built in the next few years and commercial application on a large scale may be feasible. Although considerable shale lands are privately held by oil companies, the vast majority amounting to some 80% of the total potential reserves are federally controlled and were placed under a lease moratorium in 1930. In recognition of the possible need for future shale oil production, the Department of Interior undertook to evaluate industry interest in Federal lands by offering selected tracts for lease sale in 1968. Suitable offers were not

---

[6]These data are only approximate since firms reporting to the Census Bureau are not required to provide a breakdown of research expenditures into categories such as energy research, although many firms, especially larger ones, did provide this data.

received and no sales were made. In 1971 the government initiated a second leasing effort aimed at encouraging the development of commercial operations on a controlled basis to allow an assessment of the economic cost and environmental impact of shale oil production. This prototype lease sale culminated in early 1974 in the sale of four tracts—two in Colorado and two in Utah, each about 5,000 acres. Terms were designed to encourage purchasers to proceed with serious development of commercial facilities.

Exhibit 7 shows the four major demonstration and research projects in oil shale and the participation in each. Exxon is involved in the Paraho development project as one of the 17 industry participants. Exhibit 7 also shows the eleven firms currently involved in six planned commercial oil shale projects for which environmental and exploratory work has been announced. Joint projects are typical since the estimated capital requirements for a 50,000 barrel per day plant is over $600 million and may well reach $1 billion. Furthermore, these first generation, or "pioneer" plants, based on past experience, will entail substantially greater risk than subsequent commercial operations.

Progress in the development of oil shale has not been as rapid as many had originally expected. Exxon believes that much of the delay has been caused by the uncertainty resulting from the failure of the government to develop a national energy policy which provided any reasonable basis for business planning.

Synthetics from Coal. Germany was successful in providing much of its energy requirements in World War II with synthetic fuels produced from coal. However, the technology used at that time was suited to relatively small-scale operations and the products were expensive. New coal conversion technology is probably needed to provide large volumes of coal-based synthetics economically.

The past decade has seen increasing emphasis on research and development aimed at new processes for the production of both petroleum and natural gas substitutes from coal. While government and industry have spent large sums on this work, no new process has been demonstrated for gasification, and no process has even reached the pilot plant stage in liquefaction. This emphasizes the difficulty of the job and the need for companies experienced in large and complex developments to carry on this investigation. A large amount of development work is in progress, and several processes are moving into the pilot or demonstration unit stage where investments are substantial and requirements for skilled technical people are very high.

Exhibit 8 shows more than 35 of the approximately 55 companies involved in research, pilot and demonstration projects in coal synthetics. Exhibit 9 lists 24 companies involved in planning commercial plants for producing synthetic natural gas from coal. Exhibits 8 and 9 show a large number of participants involved in synthetics projects, many of whom are not oil and gas producers. Nevertheless, Exxon believes that delays in commercial development of these potentially important energy sources also have been caused primarily by the uncertainty resulting from the lack of a national energy policy.

Solar. The term "solar energy" covers a wide variety of activity. To simplify, there are currently three major methods for converting sunlight directly into energy under active development in the U.S. These are: (1) the use of low temperature thermal collectors for heating water and for heating and cooling space in low-rise buildings; (2) use of high temperature collectors for generation of steam and electric power in large plants; (3) use of photovoltaic cells to convert sunlight directly to electric power.

Low temperature solar collectors do not involve sophisticated technology. They use solar heat directly, without converting it to electricity. Basic technology for such systems has been developed although engineering innovation is needed for cost competitiveness in most applications. Low temperature collectors can be built by present manufacturers and installers of heating and cooling equipment. It can be anticipated that a large number of manufacturing companies could compete for this market. In 1974, at least 40 companies manufactured solar collectors.

High temperature solar collectors for power generation are not expected to be operational until the late 1980's. In this type of system, solar radiation is focused by a field of reflectors onto the surface of a small boiler. The reflectors must be mounted and their orientation controlled to maintain alignment with the sun. Currently, several companies in the aerospace industry are involved in R&D of mirror assemblies, sensing, and tracting equipment. Other components, such as the boiler and turbine/generator set, would likely be built by present manufacturers of this type of equipment. It is also likely that the plants would be owned and operated by existing electric utilities.

Both low and high temperature solar collectors have the disadvantage that the sun's radiation is only intermittently available. Solar energy growth is therefore limited to certain geographical areas and by the need to develop economical storage devices that will store solar energy and release it on a continuous basis.

Photovoltaic cells or solar cells convert sunlight directly into electricity. To date much of the R&D activity on such cells has been for the aerospace program. A few companies, including an Exxon subsidiary, are beginning to develop terrestial applications for specialized purposes. However, mass production of low-cost solar cells will require major advances in technology. New manufacturing approaches and materials will have to be identified and developed to achieve such a cost reduction. It is to be expected that firms with sophisticated technological resources and the financial capability to perform long-term R&D will be involved in photovoltaic cells. Exxon conducts solar research and a subsidiary of Exxon is involved in assembling and marketing solar cells. Several oil companies have initiated R&D projects, as have firms in several other industries—communications, electronics, photographic equipment.

Solar energy is such a young field that the current list of participants probably greatly underestimates future entrants. For example, 60 companies have applied for U.S. government contracts in solar research. Certainly the structure of the solar industries cannot be forecast accurately enough at this time to presuppose the need for legislating structural limitations in the market. Indeed, this preliminary listing of participants in solar energy indicates this industry is a wide open opportunity for a great many firms, both in the oil industry and outside of it.

Summary. The data in this section show that oil companies by no means dominate other energy sources. The leading oil and gas producers are not the leading coal and uranium producers. We estimate that oil firms account for only about 18% of coal output and 32% of uranium output; and these percentages overstate the degree of overlap since many of the "oil" firms involved in coal and uranium are quite small oil and gas producers. Oil companies are pursuing R&D projects in oil shale, coal synthetics, and solar energy, but they do not dominate energy R&D. These activities are logical extensions of petroleum technology. Given the competitive structure of the existing energy markets, the structure of these new energy industries undoubtedly also will be competitive. To foreclose these new industries to any group of companies is anticompetitive. Precluding oil company participation would immediately slow down research and development, diminish the number of competitors, and ultimately reduce future production from new and existing energy sources.

III. History of Exxon's Diversification Efforts

This section will describe the reasons behind Exxon's diversification efforts in other energy activities and our record of progress.

Entry into Coal

Exxon's entry into coal was motivated by four main factors. First, in the early 1960's, Exxon, as well as others, projected that domestic production of oil and gas would peak in the 1970's. We knew that there were very substantial reserves of oil and gas located overseas; however, like others, we were becoming increasingly concerned over the national security aspects of increased imports. Thus, we concluded at this early date that there would be substantial future needs for synthetic oil and gas. It also appeared that, since coal reserves are so plentiful in this country, a high percentage of the synthetic fuels would be made from coal. Since synthetic fuels from coal would most probably be direct raw material for our refineries and chemical plants, or directly substitutable in our marketing activities, they were deemed a logical extension of our business.

For a long time, it has been technically possible to make gasoline and synthetic natural gas from coal. Nevertheless, in the early 1960's, no economic means existed for making synthetic fuels at competitive costs. Exxon anticipated that this situation would change through a combination of lower costs for synthetics through technological improvements and higher real costs for conventional oil and gas as exploration and production expanded to increased depths and to offshore areas, and as individual discoveries became smaller and thus more costly.

Second, over the years Exxon had developed considerable expertise in processing hydrocarbons in refineries. The basic chemical reactions and chemical process design concepts applicable to petroleum feedstocks have direct bearing on coal conversion to gaseous and liquid form. Much of the R&D work in refining could prove useful in developing processes for converting coal to gas or liquids.

Third, coal was a fuel with direct marketability regardless of the success or failure of synthetic research. It offered an opportunity to expand in our basic business of energy production and sale.

Fourth, Exxon had considerable experience in commercial undertakings that had similar characteristics to those to be expected with coal. These included large-scale, capital-intensive, long-lead-time ventures such as would be involved in opening modern, efficient coal mines or developing coal synthetics plants. Our oil and gas experience also well-equipped us in land and resource management. In short, we concluded that we had the capabilities required to participate effectively in these energy fuels.

Exxon's Record in Coal Mining. An important question which had to be answered before we made a decision to enter either the coal or uranium business was the availability of resources. In the case of coal, our studies utilized reports published by the United States Geological Survey and the Bureau of Mines. Both of these organizations were estimating economically recoverable domestic coal reserves, at their current costs and technology, in the range of 200 billion tons.[7] When compared to annual coal production of about half a billion tons, these reserves represented over 400 years' supply and were very substantial by any standards. Our studies indicated that of this amount of potential reserves, approximately 65% were not owned or under lease by any company then producing coal. This information and investment opportunity was available to any company with the initiative to investigate and the willingness to take the risk inherent in entering a new business.

Exxon's coal activities are the responsibility of a subsidiary, The Carter Oil Company. The purchase of undeveloped coal reserves began in 1965. Once the general availability of leaseable coal was established, the company still had to conduct a substantial amount of core hole drilling and testing to define the exact location, extent and quality of these deposits. Considerable effort was also spent to assemble mineable blocks of reserves to develop. During the next several years rights were obtained to enough reserves to support commercial operations in Illinois. We also began obtaining coal leases in the West. Exxon's subsidiaries as of year end 1974 had a total of about 7.5 billion tons of coal reserves, about 4.1 billion of this east of the Mississippi River and 3.4 billion tons in the West, mainly Wyoming and North Dakota. This is about 3.0% of the 250 billion tons of economically recoverable coal reserves in the U.S.

Since commercially viable coal synthetics operations were not projected to be developed for several years after initial purchase of coal reserves, a logical move was to start mining and selling coal as a boiler fuel. This would not only generate income on the coal investments, but would also provide the necessary operating experience in coal mining that would ultimately be needed to support the raw material production for coal synthetics operations.

Coal marketing activities began in 1967 and have been continuous since that date. Initially, coal from the Illinois reserves offered the best prospects for near-term marketing because of its proximity to major markets. In 1968, a sales contract was negotiated with Commonwealth Edison of Chicago, and Monterey Coal Company, a Carter subsidiary, was formed and began development of its first mine. Located in southern Illinois near Carlinville, this mine commenced production in late 1970. The mine

[7]Current estimates exceed 250 billion tons. See footnote at section, "Extent of Oil Company Diversification" early in this chapter.

employs about 500 people, has a maximum capacity of about three million tons of coal per year, and is one of the largest and most modern underground mines in the country. Through the experience gained in this initial operation, the managerial and operating capability to permit expanding coal operations is being developed.

In May 1974, the development of a second underground mine in Illinois was announced; construction began in the fall of 1974. This mine is scheduled for a late 1976 opening. It is designed for 3.6 million tons/year capacity and would employ 650 people when full capacity is reached in 1979. The Public Service Company of Indiana has contracted to buy this coal.

At the present time, additional sales opportunities for Illinois coal are being limited by air pollution regulations due to its sulfur content of three to four percent. As commercial equipment becomes available for removing sulfur from coal, either directly, during combustion, or from smokestacks, after combustion, the prospects for marketing our Illinois coal will improve greatly.

In May 1974, The Carter Oil Company signed an agreement with Columbia Coal Gasification Corporation which exchanged a 50% interest in certain Illinois lands containing coal reserves for a 50% interest in a like amount of Columbia's West Virginia land. This agreement allows the immediate development of the lower sulfur West Virginia reserves and commits the high sulfur Illinois coal involved to a possible gasification project when the economics and technology are commercially demonstrated. The Monterey Coal Company is to develop and operate two underground West Virginia mines expected to produce 3-4 million tons/year for 30 years at full capacity. The mines will employ 1,600 people. The first mine is scheduled to open in 1977. Each company will market its own share of the coal.

The Carter Mining Company, another subsidiary of the Carter Oil Company, is developing the company's western coal reserves. In January 1974, Carter announced the development of a surface mine in Wyoming to open in 1976 and designed for a production level of 12 million tons/year by 1980. It will employ 300 people. Although land reclamation will assure negligible long-term environmental impact, and despite a Final Environmental Impact Statement issued by the Department of the Interior in October, 1974, progress on this mine (and on three others in the area) has been restricted as a result of a court injunction in January, 1975, arising from a suit brought by the Sierra Club and others against the Department of Interior. The injunction prohibits the Secretary of the Interior from issuing mining permits and approving mining plans, pending what we consider an extraordinary regional environmental statement. In our case, this means that it is still undetermined when the Federal government will be able to approve mining plans filed with the U.S.G.S. in April 1973.

Ironically, the western coal reserves, development of which is held up by the environmental suit, have a lower sulfur content than Illinois coal and can more easily meet environmental regulations. This favorable aspect of western coal has helped offset the disadvantages of its lower BTU content and higher transportation costs to major consumption centers.

If there are further injunctions and court actions in connection with the Sierra Club suit, production of western coal reserves could not even begin until nearly 1980. Such a delay would create serious fuel crises by 1980 for 10 utilities serving a 13-state area. In our own case, production of 5 million tons/year from Carter's proposed first western mine has been sold to Indiana and Michigan Electric Company, a subsidiary of American Electric Power, and about 2 million tons/year has been sold to Omaha Public Power District.

Exxon projections indicate that, even with the originally planned rapid development of the low sulfur western coal reserves, it will not be possible for utilities to comply now or even by 1980 with the Clean Air Laws in effect. Thus, in all probability, the utilities affected by the Sierra Club suit would be forced to switch to low sulfur heavy fuel oil to meet their energy needs. This would increase oil imports in 1980 by over 550 thousand barrels per day and, therefore, increase the nation's reliance upon the imports the nation is striving to reduce.

In sum, Exxon and its subsidiaries have made effective progress in entering the coal business. We have actively sought to develop and market our coal reserves. From 1967 to 1974 we made over 500 sales contacts and 19 formal bid proposals and currently have 4 sales contracts. We continue to negotiate sales contracts which could lead to the opening of additional mines. We hope to have 8 mines operating by 1985 with a total output approaching 50 million tons per year, which will be about 5% of total coal output. This represents the maximum rate of growth we believe to be feasible. Total investment costs for the 8 mines will amount to approximately $500 million.

Exxon's Record in Coal Research and Synthetics. Exxon and our affiliate Exxon Research & Engineering Company have had an active research program in coal and synthetic fuels since 1966. This research has produced promising processes for gasifying and for liquefying coal, each of which has been tested in small pilot plants. Exxon has spent about $35 million in coal synthetics research through 1974 and estimates 1975 expenditures at $16 million. Potential coal liquefaction pilot plants, commercial coal gasification processes, plant locations, and markets are being evaluated. A coal gasification plant to produce 250 million cubic feet/day in the 1980's would require $850 million to $1 billion dollars' investment.

In research related to the direct use of high-sulfur coal, Exxon has made substantial progress in developing a new fuel gas desulfurization process for large power and industrial plants. The process is now ready to be demonstrated commercially. This development program began in 1967 and has been supported by Exxon, by a major builder of power plants, and by a number of electric utilities. Exxon's share of the project amounts to about $6 million.

Another research program aimed at developing a new way to burn high-sulfur coal and remove sulfur in the combustion process is being conducted by Exxon under a contract with the Environmental Protection Agency. This program is now in the pilot plant stage. Exxon's share of this project cost is about $2.8 million to date. Exxon has also conducted research on nitrogen oxides control for coal combustion both under government contract (about $1.6 million in government funds) and independently (about $1.5 million in company funds for nitrogen oxides control of all stationary sources, not just coal).

Exxon's Entry into Nuclear Energy

Exxon's entry into nuclear energy was based upon the following considerations:

First, our projections of energy supply/demand showed that use of electricity was to grow about twice as fast as the demand for total energy. It appeared to us that nuclear power would play an increasingly important role in meeting electric utility demand growth. Nuclear power generation is now typically projected to grow from about 8.0% of U.S. electric energy demand in 1975 to about 50% in 1990.

Second, Exxon's accumulated oil and gas exploration experience and skills and the nature of uranium ore occurrence made uranium exploration a logical extension of our historic activity. Uranium is difficult to find and, like oil and gas, has a high exploration risk. Most of the known uranium deposits in the U.S, occur in sedimentary rocks. Since oil and gas exploration requires understanding of similar geologic environments, we had a great deal of geological expertise which could be applied to uranium exploration. Also, Exxon had an extensive library of geological and geophysical data that had not yet been examined with the objective of locating uranium deposits. It seemed possible that such rock samples and detailed geological information could be reexamined for guides to locating uranium deposits. Also, Exxon held mineral leases which covered not only oil and gas but also other minerals, including uranium.

And third, the uranium exploration, production (mining and milling), and conversion to nuclear fuel are capital-intensive and technology-intensive activities with long lead times like oil production and refining. Given these considerations, expansion into uranium offered a business opportunity that matched our capabilities well and offered a reasonable profit opportunity commensurate with the risks involved.

Exxon's Record in Nuclear Energy

The nuclear fuel cycle consists of several interrelated sectors—uranium exploration, mining, and milling; uranium conversion and enrichment; fabrication of enriched uranium into nuclear fuel assemblies; chemical reprocessing of spent fuel; and safe storage and disposal of nuclear wastes.

Exploration, Mining, and Milling. Exxon initiated its uranium exploration program in the United States in 1966. To date, we have made two uranium discoveries that have been brought into production and two others that are in varying stages of evaluation. The first commercial discovery was made in 1967 in South Texas and is a relatively minor deposit. The second discovery was made in 1968 in eastern Wyoming and is a much larger deposit having substantial proven commercial potential. The Company's petroleum activities played a key role in both of these discoveries. The Texas discovery is located on a lease which was originally obtained as a petroleum prospect. The discovery in Wyoming resulted, in part, from information gained during geophysical exploration for hydrocarbons.

Initially, operation of the South Texas property was contracted to a third party. Production from this discovery was initiated in 1970 and temporarily halted in 1972 when the contractor terminated his agreement with Exxon. As a part of the settlement, Exxon acquired a nearby uranium mill with a capacity of 650 tons per day. Present plans call for the mine and mill to reopen in 1978.

Mining feasibility studies on the Wyoming discovery were completed in 1969 and in this case we decided to conduct our own mining and milling operations. Production from an open pit and associated mill began late in 1972 and was 2.1 million pounds in 1974 or about 9% of total U.S. output.

In late 1973, development of an underground mine at the Wyoming site was started. Production from the underground mine is scheduled to start in 1977, at which time the Wyoming uranium production will approximately double from current levels.

We estimate that Exxon currently has about 6% of the uranium reserves in the U.S. Those reserves which have been assessed as commercially viable are already committed under contract to the industry. Exxon's current activities in exploration are, we believe, among the more substantial in the industry. We also believe, however, that substantial on-going commitments of funds and technology by many companies will be required to find sufficient reserves to meet the nation's needs. Legislative proposals which restrict oil company participation in these activities would be counterproductive in this regard.

Nuclear Fuel Cycle Activities in Addition to Uranium Mining and Milling. In addition to its uranium mining and milling activities, Exxon is engaged in the marketing of uranium and the design, fabrication and sale of nuclear fuel assemblies in the utility industry. In addition, it provides a range of fuel management and engineering services to that industry. Exxon intends to respond to a recent ERDA request for a proposal to design, construct and operate a privately funded centrifuge enrichment plant. It is also carrying forward plant design and commercial planning for future entry into the nuclear fuel reprocessing sector. Responsibility for these activities rests with Exxon Nuclear, Inc., a wholly owned and separately managed affiliate of Exxon Corporation. Exxon Nuclear will submit additional information concerning its nuclear activities for the committee records.

Laser Fusion. Exxon Research and Engineering Company is one of the sponsors of a program at the University of Rochester to study the feasibility of laser ignited fusion of light atoms (e.g. deuterium, tritium) for the economical generation of power. The program was initiated in 1972. Out of a total estimated program cost of $5.8 million through August 1975, Exxon Research and Engineering Company will have contributed about $917,000. This includes the cost of Exxon scientists on direct loan to the University.

Exxon Research and Engineering Company's objectives in supporting this program include the development of technological expertise in lasers and laser fusion physics and the identification of potential business opportunities in the fusion fuel cycle. At least 20 to 30 years of R&D will probably be required before laser fusion can even reach the pilot plant stage.

Exxon's Other Diversification Efforts

Oil Shale. Exxon's oil shale activities have been relatively limited During the early 1960's, Exxon acquired a number of relatively small tracts of patented land and mining claims in the oil shale area of Colorado. These holdings, however, are widely scattered and need to be consolidated to form mineable blocks. Also, in the mid-1960's, Exxon was one of 6 participants in an experimental oil shale research project at the Bureau of Mines Research Center in Anvil Points, Colorado, and is one of 17 current industrial participants supporting the Paraho Development Corporation's demonstration plant for their retorting process. Exxon's expenditures on oil shale to date total about $16 million. Of this, $8.8 million went to acquire oil shale reserves in the early 1960's, $4.9 million has been spent on research, and $2.8 million on core drilling, administrative expenses, and the like. We are still hopeful of competing in an oil shale industry as and when it emerges. Exxon believes that the lack of a government policy concerning this potentially significant energy source has been a major factor in the delay of its commercial development.

Solar. Exxon has been investigating commercial uses of solar energy since 1970, when a research program was initiated to develop advanced low-cost photovoltaic devices. For the past two and a half years, this research work has been supplemented by a solar device assembly and marketing effort carried out by Solar Power Corporation, a subsidiary of Exxon Enterprises, which is an affiliate responsible for certain activities outside of oil and chemicals. Most of SPC's solar cells are currently made for use in remote locations such as microwave transmitters, ocean buoys, etc. This commercial activity is helping to determine the size and characteristics of the market for terrestial photovoltaic devices. Exxon is also conducting research in solar space heating and cooling with the goal of developing a lower cost solar thermal collector for residential and commercial application. Exxon has spent over $3 million on solar research through 1974.

Batteries and Fuel Cells. Recognizing the increasing electrification of the energy economy, Exxon has been carrying out activities in the area of electrochemistry. Fuel cells, devices that convert chemicals to electricity, have been under study in Exxon Research & Engineering since 1960. In 1970, Exxon Enterprises entered a joint development effort which was initiated with Alsthom, a division of CGE, a French electrical equipment manufacturer. The program costs will exceed $15 million by the end of 1975. The aim of the program is to develop an efficient, inexpensive power supply for use on construction sites, as standby generators, or to power electric vehicles.

Exxon Enterprises also initiated a battery development program in 1972 based on concepts developed by Exxon Research & Engineering Company. Batteries with increased energy densities are sought as storage devices to help utilities meet peak demands and as potential power sources for electric vehicles.

Although substantial effort in terms of time and money is currently being devoted to fuel cell and battery development by Exxon Enterprises, commercialization of either is not expected to occur before 1980.

IV. Benefits of Oil Company Diversification

Exxon's diversification into alternate energy sources was motivated both by our projections that development of all domestic energy sources was needed and our recognition of the many similarities with our existing business. Examples include similarities in technology such as in synthetic oil and gas production; similarities in skills such as land and resource management in coal and uranium exploration and development; and similarities in management techniques as in the planning and development of large-scale, capital-intensive, long-lead-time ventures. This same logic, in our judgment, is applicable to any oil company.

This section clearly demonstrates that oil company participation in coal, nuclear, synthetics, and energy R&D has brought substantial benefits to the nation's economy and to the consumer in the form of increased output, increased efficiency, increased R&D, and increased competition.

Increased Output

Coal. Exxon's entry into coal has led to increased coal production as follows:

	1970	1971	1972	1973	1974
Exxon's Coal Output (Million Tons)	0.3	1.2	2.0	2.7	2.5

Our coal production is expected to increase to almost 23 million tons/year when the mines announced (described previously) are completed. Our longer range objectives are for 50 million tons a year of production in 1985. Even if this effort, which we consider to be the maximim feasible, is successful, this will be about 5% of projected total U.S. coal output in 1985.

Exhibit 11 shows the production record of seven oil companies who produced coal in 1974. The exhibit shows that, while total U.S. supplies of coal increased by 2.2%/year from 1964 to 1973, the oil companies' coal production increased by 4.2%/year. Far from restricting output in an attempt to raise prices for coal, as some have alleged, these companies in the aggregate increased output at a faster rate than the industry average.[8] Since oil companies only have about 18% of coal output, they could not seek to raise prices by restricting output since buyers could simply switch and purchase their needs from the remaining 80% or more of the market.

Uranium. Given the tremendous task of meeting future nuclear energy demand, Exxon's entry into uranium exploration and production is significant for the increased output it has effected. Exxon's uranium concentrate production record is as follows (in millions of pounds of U_3O_8):

1970	1971	1972	1973	1974
0.1	0.5	0.8	2.6	2.1

The first full year of production at the Highland, Wyoming facility was 1973. The mill was capacity tested utilizing high quality ore mined in 1972 prior to the mill startup. As a result 1973 production exceeded 1974.

Only limited statistical data are available on the five oil companies engaged in uranium production. Available data for 1972-1974 are shown in Exhibit 12 and indicate that the oil companies involved in uranium production expanded output at a slighly greater rate than the average for the industry. From 1972 to 1974, their share of industry output increased from 29.8% to 32.0%. Excluding Kerr-McGee,[9] the four oil companies more than doubled their output. Since market demand for uranium in 1985 will be almost three and one-half times larger than the 1975 level, it is clear that substantial efforts from many producers, existing and potential, to increase uranium exploration and production are essential to national interests.

Increased Productivity and Efficiency

Coal. Economies of scale and changing customer needs favor larger coal mines. A firm must have adequately large coal reserves to compete for the long term contracts desired by customers. For example, utilities want assured supplies of known quality for a major portion of the useful life of their new plants before committing the huge sums of capital for such generating facilities. The second advantage of larger mines is that unit trains

---

[8]Production data for 1974 show practically no increase in industry output due to the coal strike, and a relative decline in the output of the seven oil companies in coal. Since the strike made 1974 an abnormal year, it was not included in the trend analysis.

[9]Kerr-McGee's annual report indicated that their 1974 production was slow to recover from the effects of a 1973 strike. Based on the sales figure announced in their 1974 annual report, Kerr-McGee apparently utilized available inventory to meet demand.

which travel between the mine and generating plant confer substantial transportation cost advantages to large coal shipments from a single mine. Third, the 1969 Federal health and safety rules added significant costs to coal mining which many small mines could not absorb.

In recognizing these facts, Exxon, like others, has designed its mining operations so as to maximize output; employ optimal cost-saving technology; and utilize the latest methods for ventilation, dust suppression, machine protection, and roof support for safety.

The results of Exxon's efforts in manpower development and production technology are reflected in productivity. Our mine productivity is 23 tons/man-day compared with 17 tons/man-day for all underground mines in Illinois and 11 tons/man-day for all U.S. underground mines.

While economies of scale in coal tend to favor entry on a larger scale, it should be noted that many small companies operate successfully in the coal business. In some areas, such as Appalachia, smaller mines are profitable because proximity to markets and premium quality coal can support price levels which offset higher costs. In other instances, small mines operate successfully on the spot market where long-term contracts backed by large reserves are not involved. Even if larger coal reserves are needed for entry, coal reserves are plentiful and, in general, barriers to entry in coal are relatively low.

Uranium. Entry requirements for uranium exploration, which consist primarily of costs for acreage acquisition and shallow core hole drilling for evaluation, are relatively low. However, uranium is becoming increasingly more difficult to find and many prospects must be tested year in and year out. For example, ERDA has reported that the finding rate per foot of exploratory hole drilled deteriorated from about 13 pounds per foot in 1967/68 to less than 2 pounds per foot in recent years. Nevertheless, many companies undertake modest programs for less than $200,000 a year, although the larger domestic exploration programs in the U.S. have run as high as $10 million per year.

Mining costs depend primarily on ore body size and depth. Reserves are being discovered at increasingly greater depths and are expected to require higher investments for underground mines. To open an underground mine capable of 1,000 tons of ore per day will require $15 to $25 million in constant 1975 dollars. If a potential mining property is close to an existing mill, ore production can often be custom processed or sold to the mill operator, minimizing the producer's capital requirements. If milling facilities are not close by, a 1,000 ton/day plant would require an additional outlay of $7-14 million in 1975 dollars, and a 2,000 ton/day plant would require $11-23 million. In 1974 there were 16 operable mills with plans announced for two new plants and expansion of several existing mills.

Thus total capital investments for entry into uranium mining range from as low as $200,000/year for initial exploratory work to as high as $48 million for a mine and mill. Both small and large companies can participate in the industry and capital requirements are not a large barrier to entry. As already mentioned, over 83 companies are currently involved in uranium exploration. The fact that about 20 of these are oil companies is not surprising given the exploratory expertise and knowledge of these companies in sedimentary geology and their familiarity with high-risk investment activity. Oil companies such as Exxon have demonstrated they have the financial, technical, and managerial capability to improve the productivity and efficiency in uranium mining and milling. During 1973, Exxon's Highland, Wyoming open pit uranium mine averaged 1.75 tons of raw ore per man-hour versus the average Wyoming industry rate of 0.83 tons per man-hour.[10] Productivity at the mine was the highest in the state.

In sum, the coal and uranium industries are so structured that both large and small firms can participate. While efficiency often dictates entry on a fairly large scale, barriers to entry are not high in either industry.

Increased R&D

Technological progress is one of the main stimulants to productivity and economic growth. It creates new products and new processes and reduces costs. The petroleum industry is well-known for its technological advances and cost-saving innovations in petroleum exploration, production, refining, and transportation, such as are documented in our January 1975 statement to the Senate Antitrust and Monopoly Subcommittee. Oil companies have brought the same expertise and propensity to innovate to the coal, nuclear, synthetics, and solar industries and can be expected to continue to do so unless prevented from doing so by anticompetitive legislation.

Until recently the coal industry has not undertaken extensive R&D ventures. For the past several decades, coal has been a declining industry, steadily losing its markets to other energy sources. U.S. coal production was no greater in 1966 than in 1946. Few firms in the industry were profitable enough to conduct R&D. In addition, many coal companies were and are too small to conduct R&D programs. Most coal research has been sponsored by the Bureau of Mines and the Office of Coal Research of the

---

[10]Calculated from data in the Wyoming State Mine Inspectors Report for 1973 which reported man-hours and tons produced by individual companies.

Department of Interior, and has involved technology to improve mining techniques, ventilation, safety, and other traditional concerns. As the size of coal mines increases and as new companies with much technical expertise invest in coal facilities, more coal R&D will be undertaken by the private sector. In addition, the development of coal synthetics and the passage of environmental laws has spurred R&D in new areas. For example, from 1972 to 1974, private R&D expenditures on coal by all industries nearly tripled (from $38 million to $103 million). From 1972 to 1974, Exxon's total R&D expenditures in coal rose more than fivefold from $2.7 million to $14.7 million. The 1975 estimate is $17.8 million. Over 90% of Exxon's 1974-1975 coal expenditures were company funded.

Exhibits 7, 8, and 9 have already shown the number of companies involved in R&D of coal synthetics and oil shale, including the number of oil companies. Oil firms' progressiveness in synthetics R&D is also evident in statistics collected on the number of patents granted in the U.S. From 1964 to 1974, petroleum firms were granted 49 patents covering synthetic gas and oil from coal (19 to Exxon, 11 to Arco, 7 to Continental Oil, 6 to Sun Oil, 3 to Gulf Oil, 2 to Occidental, and 1 to Kerr-McGee).

Although Exxon has been increasing expenditures in absolute amounts in almost all phases of its total R&D budget, it has been spending increasingly larger amounts both absolutely and relatively for nuclear R&D. Exxon has spent well over $50 million in nuclear research through 1974 for fundamental R&D, applied development, prototype testing, and pilot plant site operations. Exxon's estimated R&D expenditures for 1975 exceed $11 million.

There is little public data available on private R&D expenditures in solar energy. Exhibit 10 showed a number of firms who are now conducting research in the area of photovoltaic cells. These firms are from a wide variety of industries—oil, communications, computer electronics, photography, as well as research companies. The U.S. government is giving solar R&D a high priority. The Energy Research and Development Administration (ERDA) has recently started a major research program with Jet Propulsion Laboratories as project manager. About 60 companies have applied for contracts so far. Only by bringing to bear the skills and imagination of private enterprise can progress be made in this area.

Increased Competition

Clearly Exxon's entry into coal and uranium has provided an increase in the number of competitors in these fuels and an increase in production and competition. Exxon represents a new source of coal, uranium, and nuclear fuel services. Exxon's exploratory efforts added new discoveries and a new competitor to the uranium industry. Many other oil and non-oil companies have entered into uranium exploration and, if successful, they will materially expand the number of competing producers. Exxon is offering the utility industry today a realistic and dependable alternative source of nuclear fuel cores and fuel management services and a new and independent source of coal supplies.

New competitors can bring to any industry new financial, technical, and managerial talents to spur output, R&D, and competition with other firms, regardless of the method of entry. The beginning of this submission documented the fact that the top firms in oil and gas production are not the same top firms in coal and uranium production. Thus, even that amount of oil industry diversification achieved through acquisition rather than grass roots entry has not decreased the total number of energy firms competing among the top sixteen of each energy source and has not increased energy concentration levels. Rather, oil industry investment in other energy fields has stimulated competition by increasing output, improving productivity, advancing R&D, and bringing new financial sources and managerial expertise to these areas.

V. Summary and Conclusions

There is no monopoly power or threat of monopoly power by any oil company, or group of oil companies, in any individual fuel sector or in total energy. Statistics presented in this paper as well as in the Federal Trade Commission study on energy concentration ratios and oil company diversification do "...not provide any positive support for the proposal that petroleum companies be banned from acquiring coal or uranium companies; nor does it suggest that petroleum companies be banned from acquiring coal or uranium reserves."[11] The facts show that the top firms in oil and gas production are not the same top firms in coal and uranium.

Exxon's diversification into coal and uranium was prompted by an early assessment that all available sources of energy would have to be mobilized to keep pace with energy demand and that Exxon had strengths to develop efficiently and compete effectively in these energy supplies. After 10 years of effort, Exxon has one operating coal mine, has started construction on two more, and two others have been announced. We have one uranium mine and mill in operation, a second mine and mill scheduled to resume

[11] "Concentration Levels and Trends in the Energy Sector of the U.S. Economy," Staff Report to the Federal Trade Commission, January 1974, pg. 261.

production in 1978, and another underground mine under construction that will increase producing capacity at our Wyoming facility. Exxon's diversification has clearly been beneficial to the U.S. economy and the consumer. It has led to increased output, improved productivity, increased R&D, and increased competition. While oil companies have diversified into coal and uranium for their own reasons based on their own independent assessments of their capabilities and opportunities, we believe that their entry in the aggregate has had the same beneficial results.

Proposed legislation which would restrict oil company diversification would protect firms in coal, nuclear, synthetics and solar energy from a large number of present and potential competitors. Moreover, such legislative proposals would discourage existing firms in coal or uranium from entering the oil and gas business, and would discourage non-energy companies from entering into any fuel sector because of the uncertain outlook for the future which such legislation would create. As an example, Pittston Coal Company, the fifth largest coal producer in 1974, is seeking to build a large oil refinery in Maine. This new entry into oil refining would be prohibited by some legislative proposals.

Current antitrust laws such as the Sherman Act and Clayton Act ensure continued competition in energy markets without creating arbitrary limits on entrants and competitors, and are effective safeguards of competition. Moreover, the ink is barely dry on new Federal legislation—the Federal Trade Commission Improvements Act and the Antitrust Procedures and Penalties Act—both intended to strengthen the basic antitrust laws.

Proposals which restrict oil company diversification are counter to the nation's goals of increased energy independence through the development of increased domestic coal, nuclear, synthetics, and solar energy. These proposals would deprive these sectors of new sources of capital, technology and management skills. In the final analysis, those who would be hurt by such proposals are not primarily the oil companies but energy consumers and the nation as a whole.

The United States is fortunate in being endowed with a tremendous energy potential. This large, untapped potential can and will be developed by private enterprise, given the opportunity. More than anything else, the solution to our energy problems involves the development of new energy sources and technologies. This is precisely what the oil industry has been doing year in and year out over most of its history. Competition already assures that energy resources are being developed efficiently. Such competition will continue to benefit the consumer as long as artificial restraints on firms' growth and opportunities are not enacted.

EXHIBIT 1

TOP 16 COMPANIES IN EACH ENERGY SOURCE 1974 PRODUCTION

	Crude Oil & Natural Gas[1]	Coal[2]	Uranium[3]
1.	Exxon	Peabody	Anaconda
2.	Texaco	Conoco	Kerr-McGee
3.	Std. of Indiana	Occidental	Utah International
4.	Shell	Amax	Un. Nuclear
5.	Gulf	Pittston	Exxon
6.	Mobil	U.S. Steel	Union Carbide
7.	Arco	Ashland/Hunt	Rio Algom
8.	Std. of Calif.	Bethlehem	Cotter
9.	Union	N. American	Conoco
10.	Getty/Skelly	Peter Kiewit	Dawn
11.	Phillips	Sohio	Pioneer
12.	Sun	Eastern Asso.	Homestake
13.	Cities Service	Westmoreland	W. Nuclear
14.	Conoco	Gulf	Atlas
15.	Tenneco	Utah Int'l.	Federal Resources
16.	Marathon	Am. Electric Power Co.	Getty/Skelly

[1]Source: Annual Reports and Statistical Supplements.

[2]Source: Keystone News Bulletin.

[3]Source: Exxon estimate based upon ERDA mill production figures by state, Annual Reports, press releases, and mill capacities reported by ERDA.

EXHIBIT 2

TOP 16 ENERGY COMPANIES AND RANKING IN INDIVIDUAL FUELS[1] 1974 PRODUCTION

	Energy (BTU Basis)	Crude Oil & Natural Gas	Coal	Uranium
Exxon	1	1	*	5
Texaco	2	2	—	—
Std. of Indiana	3	3	—	—
Shell	4	4	—	—
Gulf	5	5	14	—
Conoco	6	14	2	9
Mobil	7	6	—	—
Arco	8	7	—	—
Std. of Calif.	9	8	—	—
Peabody	10	—	1	—
Getty/Skelly	11	10	—	16
Union	12	9	—	—
Phillips	13	11	—	—
Sun	14	12	—	—
Cities Service	15	13	—	—
Kerr-McGee	16	*	—	2

[1]Source: Same as in Exhibit 1.

Dash (—) indicates no production at all.

*indicates some production, but not in top 16. Kerr-McGee ranks about 25th in oil and gas production. Exxon is 37th in coal.

EXHIBIT 3

OIL COMPANY PARTICIPATION IN COAL PRODUCTION 1974

	Million Tons	% of U.S. Output	Rank
Continental Oil (Consolidation Coal)	51.8	8.6	2
Occidental Petroleum (Island Creek)	20.8	3.5	3
Ashland-Hunt (Arch)	13.9	2.3	8
Sohio (Big Ben)	9.5	1.6	10
Gulf Oil (Pittsburg & Midway)	7.5	1.3	13
Exxon	2.5	0.4	37
Belco (Hawley Fuels)	1.3	0.2	61
Total of Oil Companies	107.3	17.9	
Total U.S. Coal Output	601.0		

Source: Keystone News Bulletin, Volume 33, Nos. 2 & 3, February 26 & March 25, 1975.

EXHIBIT 4

1974 U.S. COAL RESERVES AND PRODUCTION, LOWER 48 STATES

	Reserves			1974 Production		
Oil Companies	**Rank**	**Billion Tons**	**% Of Total**	**Rank**	**Million Tons**	**% Of Industry**
Continental Oil (Consolidation)	1	13.5	5.4	2	51.8	8.6
Exxon	5	7.5	3.0	—	2.5	0.4
Occidental Petroleum (Island Creek)	8	3.5	1.4	3	20.8	3.5
Gulf Oil (Pittsburgh & Midway)	10	2.6	1.0	13	7.5	1.3
Atlantic Richfield	12	2.2	0.9	—	0	—
Texaco	14	1.9	0.8	—	0	—
Kerr McGee	15	1.9	0.8	—	0	—
Sun Oil	—	1.1	0.4	—	0	—
Shell	—	1.0	0.4	—	0	—
Tenneco	—	0.9	0.4	—	0	—
Ashland-Hunt (Arch Minerals)	—	0.9	0.4	8	12.5	2.1
Sohio (Old Ben)	—	0.8	0.3	10	9.5	1.6
Belco (Hawley Fuels)	—	0.3	0.1	—	1.3	0.2
Total Listed Companies		38.1	15.2		105.9	17.7
Others		2.0	0.8		n.a.*	n.a.
Total		40.1	16.0		n.a.	n.a.
Non-oil Companies						
Union Pacific	2	10.0	4.0	—	0	—
Burlington Northern	3	9.0	3.6	—	0	—
Kennecott (Peabody)	4	8.9	3.6	1	68.1	11.3
North American Coal	6	5.0	2.0	9	9.8	1.6
Amax	7	4.9	2.0	4	19.9	3.3
U.S. Steel	9	3.0	1.2	6	16.4	2.7
Eastern Gas & Fuel	11	2.6	1.0	11	7.7	1.3
Bethlehem Steel	13	1.9	0.8	7	13.3	2.2
Utah International	—	1.5	0.6	15	7.0	1.2
Pittston	—	1.5	0.6	5	17.4	2.9
Westmoreland	—	1.2	0.5	12	7.6	1.3
General Dynamics	—	0.8	0.3	14	7.0	1.2
Total Listed Companies		50.3	20.1		174.2	29.0
Others		9.6	3.9		n.a.	n.a.
Total		59.9	24.0		n.a.	n.a.
Total Industry		100.0	40.0		601.0	100.0
Open Acreage		150.0	60.0			
Total Lower 48 States		250.0	100.0			

*Not available.

Source: Exxon's estimates from Annual Reports and press releases. Total coal reserves of 250.0 billion tons is Exxon's estimate of the recoverable portion of the demonstrated resource base compiled by the Bureau of Mines. It should be recognized that these estimates of coal reserves by company are only approximations; there are no doubt instances where our estimates could deviate significantly from the appraisals that individual companies have of their own reserves.

EXHIBIT 5

OIL COMPANY PARTICIPATION IN URANIUM MILL PRODUCTION 1974

	Million Pounds	% of Total	Rank
Kerr-McGee	3.5	15.2	2
Exxon	2.1	9.1	5
Continental	1.0	4.3	9
Pioneer	0.6	2.6	11
Getty/Skelly	0.2	0.8	16
Total of Oil Companies	7.4	32.0	
Total of Other Companies	15.7	68.0	
Total U.S. Mill Output	23.1	100.0	

Source: Exxon estimates based upon U.S. Government (ERDA) mill production figures for New Mexico, Wyoming and Other U.S.; annual reports and press releases; use of mill capacities reported by ERDA.

EXHIBIT 6

INDUSTRY R&D EXPENDITURES FOR ENERGY 1973

	(Millions of $)	%of Total
Petroleum refining & extraction	325	37.2
Electrical equipment & communications	243	27.8
Aircraft & missiles	65	7.4
Chemicals & allied products	61	7.0
Non-manufacturing industries	47	5.4
Primary metals	15	1.7
Scientific instruments	13	1.5
Machinery	11	1.3
Other	95	10.7
Total	875	100.0
Company funds	593	67.7
Federal funds	282	32.3

Source: National Science Foundation, "Science Resources Studies Highlights," December 4, 1974, NSF 74-319.

EXHIBIT 7

COMPANIES IN OIL SHALE SYNTHETICS AS OF MARCH 1975

	Estimated Project Cost
Demonstration, Pilot or Research Projects	
Occidental Oil Shale, Inc.	Over $1 million
State of Colorado, D.O.I., 17 industry participants	$735,000
Institute of Gas Technology, American Gas Association	$300,000
Paraho Development Corporation (with 17 industry participants)	$7.5 million
Commercial Plant Proposals (Exploratory Work Underway)	
ARCO, The Oil Shale Corporation (TOSCO), Ashland Oil, Shell Oil	$600 million
Gulf Oil and Standard of Indiana	$600 million
Sun Oil and Phillips	$600 million
Superior Oil	$300 million
TOSCO	N.A.
Union Oil	$750 million
Sun, Phillips, Sohio	$600 million

Source: *Synthetic Fuels*, Vol. 12, #, March 1975, quarterly report by Cameron Engineers, Inc. (In many cases, project costs are from initial press releases and thus may be significantly lower than current estimates.)

EXHIBIT 8

SYNTHETIC FUELS FROM COAL
DEMONSTRATION, PILOT & RESEARCH PROJECTS

1. BuMines, Foster Wheeler & Bethlehem Steel
2. BuMines, Consolidated Coal (Conoco)
3. BuMines, Union Pacific Railroad
4. Catalysts & Chemicals, Inc.
5. COGAS (Consolidated Natural Gas, FMC Corporation, Panhandle Eastern Pipeline, Republic Steel, Tennessee Gas Pipeline)
6. Commonwealth Edison
7. Conoco Methanation Company
8. Conoco & 13 other U.S. companies
9. Electric Power Research Institute & the Southern Company
10. El Paso Natural Gas
11. EPA & Applied Technology Corporation
12. Exxon
13. General Electric
14. Gulf R&D
15. Hydrocarbon Research (sponsored by Ashland, ARCO, Standard of Indiana, Exxon Research & Engineering)
16. Institute of Gas Technology
17. Occidental (Island Creek Coal)
18. Sohio (Old Ben Coal Corporation)
19. Universal Oil Products
20. Wheelabrator-Frye, Inc.

Many Office of Coal Research Projects with many industry participants

Source: Same as Exhibit 7

EXHIBIT 9

SYNTHETIC FUELS FROM COAL
COMMERCIAL PROJECTS (IN PLANNING STAGES)

	Estimated Project Cost
Cities Service & N. Natural Gas Company	N.A.
Colorado Interstate Gas Company	N.A.
El Paso Natural Gas Company	$600 million
Exxon	$250 million
Illinois Coal Gasification Group (8 public utilities)	N.A.
American Natural Gas	$800 million
Peoples Gas Company	N.A.
Panhandle Eastern Pipeline & Peabody Coal	$600 million
Texaco	N.A.
Texas Eastern Transmission, Pacific Lighting Corporation, WESCO	$450 million
Texas Gas Transmission Corporation	$200 million
Transco/Mobil	N.A.

Source: Same as Exhibit 7, except Exxon's estimated project cost has been updated.

EXHIBIT 10

COMPANIES PARTICIPATING IN PHOTOVOLTAIC (SOLAR) CELLS

1. Solarex Corporation; Rockville, Md.
   Most of activity on terrestial solar cells.
2. Solar Power Corporation; Wakefield, Mass.
   Most of activity on terrestial solar cells. (Owned by Exxon Enterprises)
3. Spectrolab; Sylmar, Calif.
   Active both on terrestial and space solar cells.
4. Centralab; San Francisco area.
   Active on space solar cells.
5. Mobil Oil and Tyco Labs; Waltham, Mass.
   Joint venture company trying to develop low cost silicon ribbon cell.
6. Solar Energy Systems; near University of Delaware.
   Supported to some extent by Shell Oil Co. Has research program on cadmium sulfide cell.
7. Bell Laboratories; Murray Hill, New Jersey.
   Developed silicon solar cell 20 years ago. Has advanced semiconductor laboratory.
8. IBM; Yorktown Heights, N.Y.
   Has large research program on gallium arsenide cell.
9. Eastman Kodak; Rochester, N.Y.
   Has research program of unknown size. May be investigating light sensitive organic compounds as semiconductors.
10. Texas Instruments; Texas.
    Has research program on silicon cell.
11. RCA; Princeton, N.J.
    Has research program of unknown size.
12. Varian; Palo Alto, Calif.
    Has research program on gallium arsenide cell.
13. Innotech; Connecticut.
    Has research program on glass-like materials for use in solar cells.

EXHIBIT 11

OIL FIRM COAL PRODUCTION
Millions of Tons

Coal Company	Parent Co.	1964	1965	1966	1967	1968	1969	1970	1971	1972	1973	1974
Consolidation & Affiliated Cos.	Continental	45.4	48.6	51.4*	56.5	59.9	60.9	64.1	54.8	64.9	60.5	51.8
Island Creek	Occidental	21.2	20.6	23.7	25.9	25.9*	30.3	29.7	22.9	22.6	22.9	20.8
Old Ben	Std. of Ohio	5.1	6.3	9.9	10.3	9.9*	12.0	11.7	10.5	11.2	10.8	9.5
Pittsburgh & Midway	Gulf	7.1*	8.2	8.8	9.0	9.2	7.6	7.8	7.1	7.7	8.1	7.5
Hawley Fuels	Belco	1.0	2.1	1.9	1.7	1.5*	1.7	1.8	1.4	1.6	1.5	1.3
Arch Minerals	Ashland	2.3	5.1	6.8	7.5	7.0	6.8	6.3	7.2*	11.2	12.5	13.9
Monterey Coal	Exxon	—	—	—	—	—	—	0.3	1.2	2.0	2.7	2.5
Seven Company Total		82.1	90.9	102.5	110.9	113.4	119.3	121.7	105.1	121.2	119.0	107.3
Industry Total		487.0	512.1	533.9	552.6	545.2	560.5	602.9	552.2	595.4	591.7	601.0
Seven Company Output as Percent of Total		16.9	17.8	19.2	20.1	20.8	21.3	20.2	19.0	20.4	20.1	17.9

*Year acquired by oil company.

Keystone Coal Industry Manual (McGraw-Hill), various years.

Concentration Ratios in Production (%), 1974

	Top 1	Top 4	Top 8
Oil (net liquid hydrocarbons)	9	26	42
Gas	9	25	38
Coal	11	27	37
Uranium (mill production)	17	58	85
Total Energy (BTU Basis)	8	21	34
All U.S. Manufacturing Average	—	39	60

EXHIBIT 12

OIL COMPANY PARTICIPATION IN URANIUM PRODUCTION
Million Pounds U_3O_8

	1972	1973	1974
Kerr McGee	6.1	4.4	3.5
Exxon	0.8	2.6	2.1
Continental	0.2	1.1	1.0
Pioneer	0.2	0.6	0.6
Getty/Skelly	0.4	0.5	0.2
Oil Company Total	7.7	9.2	7.4
Industry Total	25.8	26.5	23.1
% Oil Company	29.8	34.7	32.0

Source: Same as Exhibit 5

16

Who Should Develop Coal? Everybody!

TESTIMONY OF C. HOWARD HARDESTY, JR., A VICE CHAIRMAN OF CONTINENTAL OIL CO., BEFORE THE SUBCOMMITTEE ON MONOPOLIES AND COMMERCIAL LAW, HOUSE JUDICIARY COMMITTEE — SEPTEMBER 11, 1975

Summary: Oil companies have served to increase investment in U.S. coal production and help toward the national goal of doubling our coal output as soon as possible. Oil companies stimulate and benefit the coal industry by bringing to bear both the ability to raise large amounts of capital efficiently and proved skills in managing large capital projects; applying its skilled manpower and technology to speed research into new fuel conversion methods, e.g., coal into gas; increasing competition. Conoco's answer to "Who should develop coal?" is "Everybody."

C. Howard Hardesty, Jr. speaks . . .

I am a Vice Chairman of Continental Oil Company, a member of Continental's Board of Directors and a member of its Management Committee. Prior to my transfer to Continental Oil in 1968, I was Executive Vice-President of Consolidation Coal Company, which I originally joined in 1963 as General Counsel.

Adequate supplies of energy in all its forms—oil, gas, nuclear, and coal, among others—are central to our national security and economic growth. The need at least to double our coal production has been identified as a key element in our national energy policy, and should not be an issue today.

The question is whether an oil company serves the national interest by assisting this vital expansion of the coal industry. Because of my association with both Continental Oil Company and Consolidation Coal Company, I am pleased to participate in these hearings.

I. BENEFITS RESULTING FROM CONTINENTAL OIL COMPANY'S ACQUISITION OF CONSOLIDATION COAL COMPANY

Because I am so strongly convinced that the association of Consol and Conoco has benefited the industry and the nation, I would like to specify those benefits and detail how it has promoted competition and stimulated coal production and important research.

Background of the Acquisition

Continental began to diversify its operations in the late 1950's and early 1960's due to the declining outlook for growth in the U.S. petroleum industry. Up to that point, Continental had been primarily a domestic oil company, but rising costs of finding and developing domestic oil, declining prices of petroleum products, and an artificially low price for natural gas under Federal Power Commission regulation limited opportunities in the domestic oil industry.

In view of these limitations in its traditional area of operations, Continental diversified along three main lines: first, since foreign oil was less expensive to find and produce than domestic oil, it began to develop overseas oil operations. Second, it moved into areas, such as petrochemicals, that were logical extensions of its oil business. And third, since demand for electricity was expected to grow very rapidly in coming years, Continental was attracted to the market for fuel to generate electricity. Continental began uranium exploration in 1967 and opened its first mining and milling complex in 1972. Following reviews by the Department of Justice and IRS, Conoco acquired Consolidation Coal Company in 1966.

Benefits of the Acquisition

The benefits which have come from Continental's participation in coal are numerous.

1. *Stepped-up capital spending.* In the five years prior to its acquisition by Continental, Consol's capital outlays for new mines and expanded

capacity averaged $13.5 million a year. In the years since the acquisition, the yearly average for new mines and expanded capacity was approximately $36.5 million a year.

2. *Increased employment.* During the post-acquisition period, employment increased significantly as a result of expansion programs. In December 1966, Consol employed 11,697 people. By the end of 1974, employment had increased 40 percent to 16,351 people. Continental has been able to apply many of its training, recruiting and organization programs in making these additions to Consol's work force.

3. *Increased coal production.* In the initial four years after acquisition, the Company's coal production increased from 51.4 million tons to 64.1 million tons (up 25 percent) as a result of the stepped-up pace of capital spending.

   Throughout the industry, deep mine production decreased from 1970 to 1974. Our decline in coal output from deep mines was only 14 percent compared with an industry-wide decline of 19 percent in the 1970-74 period despite the fact that approximately 70 percent of Consol's production comes from deep mines compared to 48 percent for the industry. A number of companies moved to counter the decline in deep mine production by expanding surface mine output. Surface mine output for the industry increased 28 percent between 1970 and 1974. This increase was largely in the West. Consol participated in this expansion to some degree. But our growth in western surface mining has been limited because of the present moratorium on leasing of Federal coal lands in the West, in effect since 1971. Consol did not own large western reserves and due to this moratorium, it has not been able to block up sufficient new reserves to develop western production.

4. *Perspective on recent price increases.* Coal prices have increased sharply since the end of price controls in April, 1974. However, these increases were required for two reasons: (a) to compensate for cost increases which had reduced Consol's earnings to about the break-even level in the early 1970's and to a loss in 1973, and (b) to provide margins commensurate with the risk and necessary scale of investment to assure new coal production. During the past two years, the coal industry and Consol have experienced sharp cost increases. Many of these increased costs arising from health, safety and environmental considerations are socially desirable, but they must be reflected in the market price for coal to justify bringing new reserves into production.

5. *Research.* Achievements in any industry research endeavor flow only

from the dedicated effort of talented and inspired men and women. It is commonly agreed that the interaction of scientists and technicians from varied disciplines and experiences brings forth new concepts and quickens technological development. This has occurred from the interaction of Continental and Consol research and engineering capabilities and facilities.

Combined Conoco and Consol expenditures on R&D in the areas of pollution-free synthetic fuels, mine health and safety, environment safeguards, operational efficiency, and improved transportation methods demonstrate the accelerated tempo of our research. Taking the areas of liquefaction, gasification, and pollution abatement as an example: $1.2 million was spent during the four-year period from 1963 to 1966. In the eight-year period, 1967 to 1974, such expenditures amounted to $9.4 million, a nearly fourfold annual increase. The scope and accomplishments of our efforts are illustrated by a brief summary of our principal programs.

(a) A feasible coal liquefaction process has been demonstrated by "Project Gasoline" at Cresap, West Virginia. This process has substantial cost advantages over earlier technology. Continental's engineers, utilizing the company's oil refining know-how, made important contributions to development of coal liquefaction in operations as well as in process design.

(b) The synergistic effect of oil-refining technology upon coal processing has been most dramatically demonstrated in our highly successful, large scale coal gasification work at two locations. At Rapid City, South Dakota, Consol's carbon dioxide acceptor process has successfully produced low BTU gas. This large pilot plant was designed by Continental's engineers who also supervised construction and assisted extensively during start-up. Because work as well as the coal liquefaction program was pursued in conjunction with the Office of Coal Research, the know-how and technology are automatically available to all firms.

In Westfield, Scotland, a second coal gasification program converted Lurgi process low-BTU gas to pipeline quality gas on a commercial scale. Conoco acted as program manager for the 16 participating U.S. companies. This program lasted 30 months and cost $6 million. In this case, not only did Continental's engineers design, construct, and operate the large scale methanation facility, but Continental's researchers gathered the design data from a small bench-scale unit in our Ponca City, Oklahoma laboratories and provided technical service assistance throughout the entire program. At the conclusion of this commercial-scale test in August, 1974, 2.5 million cubic feet per day of the synthetic gas were fed into the local gas distribution grid where it was used by several thousand Scottish consumers. This project verified a large scale methanation process and represented the final step in proving technology required for manufacturing substitute pipeline quality gas from coal.

The project at Westfield also confirmed that methanol—an easily transported, easily stored, clean burning fuel—can be produced in commercial-size quantities from coal. Consol has subsequently joined a group of 14 other United States companies to finance the development of an improved version of the Lurgi gasification system at Westfield. This work on the so-called slagging gasifier is currently in progress.

(c) Mine safety and productivity programs have been accelerated. Research expenditures in this area amounted to over $9 million since inception, including 1975 budgeted outlays. These expenditures, initiated about two years prior to the enactment of the 1969 Coal Mine Health and Safety Act, were in addition to previously cited research programs. The principal efforts have been directed to:

(1) A hydraulic transportation system to provide continuous transport of coal in coarse, aqueous slurry form from the mine face to the preparation plant. This system is designed to increase mine safety and productivity by eliminating the hazards and bottlenecks of transporting coal by traditional methods. Continental's pipeline technology and know-how have been fundamental to this research. We hope the first demonstration of this entire underground hydraulic transportation system will occur this month in Consol's Robinson Run mine.

(2) A program to remove methane gas from coal mines. Oil field drilling skills have been applied in the development of several new techniques: (a) the drilling of long horizontal holes in the coal seam to predrain methane before mining, (b) a system of sealing coal fracture systems with silica gel to reduce methane escape into underground shafts and (c) drilling wells from the surface into the cave "gob" zone (which results when coal is extracted underground). These wells have successfully

removed about one-half of the methane gas from these cave zones, thereby lessening the danger to active mining areas. These techniques demonstrate the use of oil field know-how in coal mine applications.

(3) Continental Oil Company's geophysical and geological technology has been used in the development of a seismic mine monitoring system for detection not only of roof falls, but also for location of trapped miners. An air monitoring system has recently been installed in the Loveridge mine which operated in conjunction with the seismic monitoring system. The air monitoring system consists of four major stations which monitor methane, carbon monoxide, temperature, rate of change of temperature, and air velocity. The system is now operating on a full-time basis.

(4) The first successful application of underground mine design utilizing rock mechanics principles was recently completed. Two long walls in Northern West Virginia were designed by Conoco Research personnel and successfully implemented in coordination with Consol operators.

(d) A wide range of environmental protection programs has been undertaken. Over half of all domestic coal is classified as high sulfur. In the eastern United States, where the largest markets for coal have historically been located, only about 11 percent of recoverable reserves are classified as low sulfur, and most of these low sulfur reserves are metallurgical coal, unsuitable as fuel in electricity generation. Urban areas can use high sulfur coal for electrical generation only if stack gases are scrubbed to remove sulfur dioxide. Currently available scrubbing systems have been difficult and costly to operate and maintain. Second-generation scrubbing systems are expected to reduce costs and improve efficiency and reliability. If these processes are perfected, over 80 billion tons of high sulfur, currently unusable coal can be made available to eastern utilities as an environmentally acceptable fuel to generate electric power.

Continental engineers have developed and designed a stack gas scrubbing unit which recovers elemental sulfur as a byproduct. It will permit the burning of high-sulfur coal without polluting the environment.

In summary, Continental's participation in coal has stimulated capital spending, employment, safety, coal production and research. In so doing, it has established no new or increased barriers to entry into those industries. On the contrary, Continental's entry into coal has stimulated competition through increased production and increased innovation.

II. INDUSTRY STRUCTURE AND MARKET PERFORMANCE

Critics frequently label the oil industry as monopolistic and charge that this is the reason for recent increases in oil prices. They also contend this monopolistic structure will spread to other fuel industries if oil companies begin to operate in them. Yet, all the data we have about this issue indicates just the opposite, namely the oil industry is intensely competitive.

A. Competition in the Oil Industry

All of the available facts confirm that the oil industry is highly competitive. In the face of overwhelming evidence, some industry critics have been forced to retreat on this issue. Now these critics have resorted to criticism of other aspects of the petroleum industry to justify their contention of anti-competitive behavior. Joint ventures among oil companies are now decried by critics as leading to a unique form of structural integration permitting exercise of monopoly power. Actually these joint ventures increase competition.

In sum, accepted measures of competitiveness—low to moderate concentration ratios and changing relative market shares—characterize all phases of the oil and gas industry.[1]

Technological innovation vigorous. Rapid technological change has occurred historically in the principal petroleum activities. This is characteristic of a dynamic competitive industry seeking to reduce or retard cost increases and improve product quality. These efforts are obviously necessary for success in the marketplace. Beyond this, however, the petroleum industry faces a basic problem different from many other industries. The most easily detected and closest to market petroleum reserves are developed first; so supplies developed later tend to be higher cost. Technological innovation is necessary to hold down this basic tendency toward increasing costs of oil and gas.

Over the last twenty years, oil companies have made remarkable progress in techniques of finding and developing new oil. For example:

(1) *In developing resources ever deeper in the earth.* In 1930, oil drilling had probed less than two miles under the earth's surface; now, the deepest wells go down nearly six miles in the search for oil and gas.

(2) *A far larger proportion of oil is now recovered from the reservoirs.* Twenty years ago, only about one-quarter of all the oil in place in fields which had been discovered was being recovered. Now, on average, the industry recovers about one-third of the oil in place in the nation's oil reservoirs. Under optimum conditions, using the best available technology for primary and secondary recovery programs, approximately 50% of the original oil in place can be recovered from a typical reservoir.

(3) *Oil companies have developed the ability to operate effectively in increasingly hostile regions as they seek new petroleum supplies.* Just after World War II, oil companies began drilling in the marshes and shallow water offshore in the Gulf of Mexico. Now, oil companies are drilling and producing oil under violent weather conditions in the North Sea in water depths up to 500 feet. Exploration for hydrocarbons is now underway in water over 1,000 feet deep. (Such techniques can be employed in our own Outer Continental Shelf waters, at such time as the government permits us to proceed.) Without such advances, however, present production from offshore areas—about 16% of current domestic oil output—would not have been available.

Conclusions about market performance. A factual appraisal of petroleum companies' performance indicates conclusively that a high level of competition exists. Long-term profitability is reasonable. Product quality has increased and petroleum price advances have been moderate. Entry of new firms occurred and is occurring in all functions. No artificial barriers bar new competitors, as has often been alleged. Finally, marked technological advances have provided relatively low-cost, high-quality petroleum products.

In this situation, the consumer benefits from the lowest price possible. This is the way it should be: new entrants and existing producers have to be efficient to compete. In a competitive situation such as this, an umbrella is not held over the inefficient producer.

B. Competition in the Coal Industry

Competition within the coal industry is as intense as in the oil industry. In coal production, the four largest companies accounted for only about 27% of total output and the top eight companies about 37% in 1974. Of the four largest coal producers, two were owned by companies engaged in the petroleum business (Continental and Occidental). Some people interpret this ownership as evidence of control over energy markets. In doing so, they overlook these facts. Although these two oil companies own the second and third largest coal producers, neither oil company is among the top eight crude oil producers and one (Occidental) holds only a minor position in the U.S. petroleum business. All the petroleum companies operating in coal accounted for only about 18% of total coal production in 1974. The matter of ownership by oil companies neither adds nor detracts from the fundamental fact that there is a low level of concentration in the coal industry. The truly important consideration is that the largest coal companies conduct a relatively modest part of that industry's total activity.

Petroleum companies own only a slightly larger share of total recoverable U.S. coal reserves than their modest portion of coal industry production. Based on a conservative estimate of recoverable coal reserves, the top four coal companies held about 28% of the total recoverable reserves in 1973, and the top eight held about 42%. Of the top four coal reserve owners, Continental Oil Company was the largest with 7.9% of the industry total; the other three were non-oil companies. Petroleum companies accounted for only 24% of total recoverable coal reserves in 1973. (Exhibits I and II.)

Some 40% of coal resources are estimated to be under government lands. Based on estimates of economically recoverable reserves of 150 billion tons, this would amount to 60 billion tons held by the government. When these reserves are opened up for leasing, there will be ample opportunity for new entrants into the coal industry.

Recent trends in coal industry concentration can be better understood in the light of historical perspective. Some increase in concentration of coal production occurred in the 1940-1965 period, as large coal companies acquired small coal producers. This trend sprang from the need to develop a more efficient organization of production. Small scale coal operations were becoming increasingly uneconomical because of (1) the very large mines required for electric utility generating stations, (2) the introduction of unit trains, which required a large single supply point, (3) the development of high voltage transmission lines serving mine-mouth generating plants, (4) the high cost of modern cleaning plants needed for coal supplied to public utilities, and (5) the need for increased mechanization to reduce coal mining costs.

Since the mid 1960's, new entrants into the coal industry have slowed the acquisition of small coal companies by large coal companies. Since that time, the acquisition of large coal reserves by new entrant oil and non-oil companies has introduced significant new competitive forces into the coal industry, rather than decreasing competition as is sometimes alleged.

C. Competition in the Energy Market

One issue of particular interest to this committee is the degree of competition between oil and coal and possible consequences of joint ownership of both oil and coal resources by oil companies. My feeling is that there are many benefits which derive from this association while at the same time competition is adequately preserved: first, because interchangeability among fuels is limited and secondly, because ownership of total energy sources is so diffused as to preclude any antitrust concerns.

1. *Competition Among Fuels*
   The energy concept, I believe, is a useful frame of reference. It is a concept sufficiently comprehensive to indicate the complex of relationships which must be considered in setting public policy. In a very broad sense, energy sources have some common characteristics. For example, heat and power are derived almost solely from inanimate energy sources. The concept "energy" might be compared to the concept "transportation." Transportation includes freight and passengers; it includes cars, buses, trains, water movement, and air movement. Many of these modes of transportation are not interchangeable. While the energy concept is useful in setting public policies, I don't believe it is the relevant "market" for antitrust considerations. By necessity, a legal definition needs to be reasonably precise. A "market" should be defined by sensible geographic boundaries and direct interchangeability of the product by the user. There are numerous energy products and uses. Many are not interchangeable. For example, natural gas or coal cannot be used, under present technological and price constraints, as a fuel for autos. Gasoline, conversely, cannot be utilized for heating purposes. Nuclear power is restricted largely to electric generation, while oil—a more versatile energy source—can be used for transportation, home heating and electricity generation. For these reasons, I conclude that energy is too imprecise a concept to measure competition for antitrust purposes.
   The utility market is the primary area for interfuel competition, but here the evidence indicates interfuel competition is decreasing. As a result of FPC price regulation, natural gas is simply unavailable to new utility plants and so is not available as a substitute for other fuel. The Federal Energy Agency is increasingly mandating the use of coal. For

instance, the FEA recently issued orders to 25 utilities to convert to coal; they have also notified 41 companies that new power plants must have coal burning capacity. Where these regulations apply, coal and oil no longer compete.

Exhibit 2 shows the delivered costs of oil and coal consumed by most electric utilities. It is immediately apparent that the price of the two fuels are greatly different. According to the April 1975 FPC report on fuel costs, the delivered cost of oil to electric utilities was $12.79 per barrel (208.8¢ per million BTU). On an oil equivalent basis, utilities paid4.93 per barrel for coal (80.5¢ per million BTU).

The inclusion of transportation costs should not materially change the relationship between coal and oil prices. In fact, since transportation costs have probably increased more rapidly for coal than for oil, the actual correspondence between oil and coal prices at point of production would be even less.

Both coal and oil prices have trended upward from 1969 to 1974 but the size of their increases have been markedly different and the reasons for price rises for coal and oil have been quite different.

A possible explanation for little correlation between coal and oil prices is the structure of the coal market. According to a Mitre Corporation Study for the FEA ("An Analysis of Steam Coal Sales and Purchases"), 75% of steam coal is sold under long term contract. The price of coal sold under long term contract is generally unrelated to the price of fuels in other energy markets. Contract price increases are subject to the careful scrutiny of customers. The buying power of large users (mainly utilities) tends to preclude unjustified price increases by coal producers that are not a reflection of real cost increases.

Rather than responding to higher prices of other fuels, coal price increases in recent years have been caused by rapidly escalating production costs. Factors contributing to higher costs were: (a) productivity declines caused by the Federal Coal Mine Health and Safety Act of 1969; (b) increased labor costs; (c) inflation in capital equipment costs; and (d) higher material and supply costs. Historically, return on investment in the coal industry has been very low. If coal production is to be sufficient to meet future demands, profit margins must be high enough to justify new mines. Only in the past year or two have coal prices begun to approach this level.

2. *Concentration in Fuels Industries*

A further consideration in treating all energy fuels as a single industry is that market shares are so small in this wider market as to be almost meaningless. The situation is comparable to citing an automobile manufacturer's share of the transportation business or a steel firm's share of the metals business. Exhibit IV shows concentration ratios for total energy production in the U.S. in 1974. The largest firm controls only 6.8% of the market, while the largest four firms shares only 19%. Because of end-use restrictions mentioned above, the number of customers who can choose between oil and coal is very small. Consequently, the market price for coal and oil is set by competition in two largely different markets. The price level determined by these markets would indicate to the few customers who had the choice which fuel to buy. Continental would compete in each market and would certainly permit Consolidation Coal Company to sell coal to utility companies that might have been previously buying oil. With 2% of the petroleum market, 9% of the coal market, 4% of the uranium market and less than 2% of the market for natural gas, Continental is not in a position to have any significant influence on the prevailing price levels of these fuels.

III. REQUIREMENTS TO MEET THE NATIONAL GOAL OF INCREASED COAL PRODUCTION

The primary objective of national energy policy is to provide adequate and secure energy supplies to American consumers at reasonable prices. This objective is not a partisan matter: the Administration, Congress and the public agree that we must reduce our vulnerability to foreign pressure and accelerate development of domestic energy supplies. Events of the past two years have shown that as long as the U.S. is highly dependent on foreign energy sources, it is vulnerable to supply interruptions—with adverse effects on its economy and employment—and to further arbitrary energy price increases.

To reduce over-reliance on energy imports requires a strong increase in our domestic production of energy.

A. Capital Requirements For:

1. *Increased production of domestic energy supplies.*
   In a forecast for the Federal Energy Administration Arthur D. Little, Inc. estimates more than $1 trillion (1974 dollar value) will be needed between 1974 and 1990 for domestic energy investments (Exhibit V). If even a moderate 5% inflation rate is assumed, investment could increase to $1.4 trillion by the time the outlays are actually made.

2. *Increased production of coal.*

Over this period, the coal industry may have to more than double its output; this means opening 400-500 new coal mines (Exhibit VI). The total investment needed to develop these new mines would probably be over $27 billion in 1974 dollars. This does not include any outlays for transportation facilities or the conversion of coal to gas or liquids.

Even with improved profitability within the last year internally generated funds may be insufficient to meet coal companies' capital requirements for the future if the industry seeks to double its production by 1985. There is the already high and rising cost of both new and replacement equipment required to meet health and safety, reclamation and other environmental regulations, for example. And the cost of research and development in new, more productive mining technology and alternative fuel forms from coal will continue to increase.

B. Sources of Investment Capital

It is a fair question to ask just who will develop the nation's domestic energy resources were petroleum companies not permitted to participate. Many *mining companies* (which have some applicable expertise in these areas) are already precluded for either antitrust reasons or lack of capital. For example, Kennecott Copper is being forced to divest Peabody Coal, while Amax has sold a 20% equity interest to another company (subject to Amax stockholder approval) partly to finance its coal expansion.

Electric utilities, which theoretically would be possible entrants, have a well-known scarcity of investment funds. In short, there is no waiting in line of possible entrants into the energy business.

Optimum development of coal logically would include those firms which have the skilled people with sufficient knowledge and expertise to be willing to commit the firm's capital to large scale energy development. Petroleum companies, have characteristically undertaken major investment programs. In 1974 they spent $17.4 billion, on investment in plant and equipment. Continental Oil—only the eighth largest petroleum company based on sales—invested over $490 million in the United States. Worldwide outlays for all its activities were $750 million.

A company supplying more than one fuel can sustain a steady investment program through good years and bad. Although in any given year earnings from one fuel may be poor, such a company can maintain investment because the profit results from other fuels are probably more favorable.

CONCLUSIONS

It is in the national interest to double the output of coal as soon as possible at the lowest price; thereby reducing our vulnerability to insecure foreign fuel supplies. The need for accelerated development of coal is so great that the real answer to the question of "who should develop coal" is "everybody." Coal production should be open to everybody. Oil companies can constitute a stimulating and productive new element in the industry for three reasons:

1. Oil companies have proven skills in managing large capital projects. Their financial strength can be a great asset in generating the large amounts of capital needed.
2. The petroleum industries can supply its technology and skilled manpower in research to further fuel conversion methods, e.g., coal into gas.
3. Oil company participation in the coal industry tends to increase, rather than decrease, competition.

Getting results in the national coal development program cannot come about through restrictions. By opening up the industry to all, including oil companies, every firm that can do the job will do the job; this is the fastest, most reliable way to serve the public. We see no interfuel rivalry between coal and petroleum. Furthermore, there is the already well established safeguard—antitrust laws—to protect the consumer.

To sum up, my personal experience indicates the desirability of petroleum companies taking part in the urgently needed development of this nation's coal industry to make the most of the vast potential of our nation's coal resources.

EXHIBIT 1

Derivation of Total Recoverable U.S. Coal Reserves[1]

A. Underground Coal Reserves (Mineable by Underground Mining Methods)

Billions of Tons		
Remaining Measured and Indicated Reserves[2]	Economically Available Reserves[3]	Recoverable Reserves[4]
349.1	209.2	104.6

B. Surface Coal Reserves (Mineable by Surface Mining Methods)

Recoverable Reserves (Billions of Tons)
45.0

C. Total Recoverable U.S. Coal Reserves (Billions of Tons)

1. Total Recoverable Underground Coal Reserves	104.6
2. Total Recoverable Surface Coal Reserves	45.0
Total Recoverable U.S. Coal Reserves	149.6

[1]Source: National Petroleum Council, "U.S. Energy Outlook" (December, 1972) Based on USGS Bulletin 1275.

[2]Bituminous, subbituminous, and lignite in seams of "intermediate" or greater thickness with overburden of less than 1,000 feet.

[3]Excludes lignite and "intermediate" thickness seams of bituminous and subbituminous coal.

[4]Based on 50-percent recovery of economically available reserves.

EXHIBIT 2

OWNERSHIP OF COAL RESERVES (1973)

Company	Reserves[1] (Millions of Tons)	Percent of Recoverable Reserves
1. Continental Oil Company	11,811	7.9%
2. Burlington Northern	11,400	7.6
3. Union Pacific	10,000	6.7
4. Kennecott	8,900	5.9
5. Exxon	7,000	4.7
6. North American Coal Corporation	5,000	3.3
7. American Metal Climax	4,900	3.3
8. Occidental Petroleum	3,500	2.3
9. U.S. Steel	3,000	2.0
10. Mobil Oil	3,000	2.0
11. Gulf Oil	2,600	1.7
12. Eastern Gas & Fuel Associates	2,600	1.7
13. Pacific Power & Light	2,500	1.7
14. Atlantic Richfield	2,200	1.5
15. Sun Oil	2,200	1.5
16. Texaco, Inc.	2,000	1.3
17. Bethlehem Steel	1,800	1.2
18. American Electric Power Corp.	1,500	1.0
19. Pittston Co.	1,500	1.0
20. Kerr-McGee	1,500	1.0
Total	88,911	59.3
Other Known Privately Held Coal Reserves	13,829	
Total Known Privately Held Coal Reserves	102,740	
Total Recoverable Coal Reserves[2]	150,000	

[1]Sources: 1974 *Keystone Coal Industry Manual,* p. 621, Company annual reports and 10-K reports, and *Forbes,* November 15, 1974, p. 67.

[2]Source: Based on USGS Bulletin 1275.

EXHIBIT 2 (Cont'd)

PETROLEUM COMPANY OWNERSHIP OF COAL RESERVES (1973)

Company	Reserves (Millions of Tons)	Percent of Recoverable Reserves
1. Continental Oil Company	11,811	7.9%
2. Exxon	7,000	4.7
3. Occidental Petroleum	3,500	2.3
4. Mobil Oil	3,000	2.0
5. Gulf Oil	2,600	1.7
6. Atlantic Richfield	2,200	1.5
7. Sun Oil	2,200	1.5
8. Texaco, Inc.	2,000	1.3
9. Kerr-McGee	1,500	1.0
Total	35,811	23.9

EXHIBIT 3

YEARLY AVERAGE COST OF FUELS BURNED BY ELECTRIC UTILITIES (CENTS/MILLION BTU's)

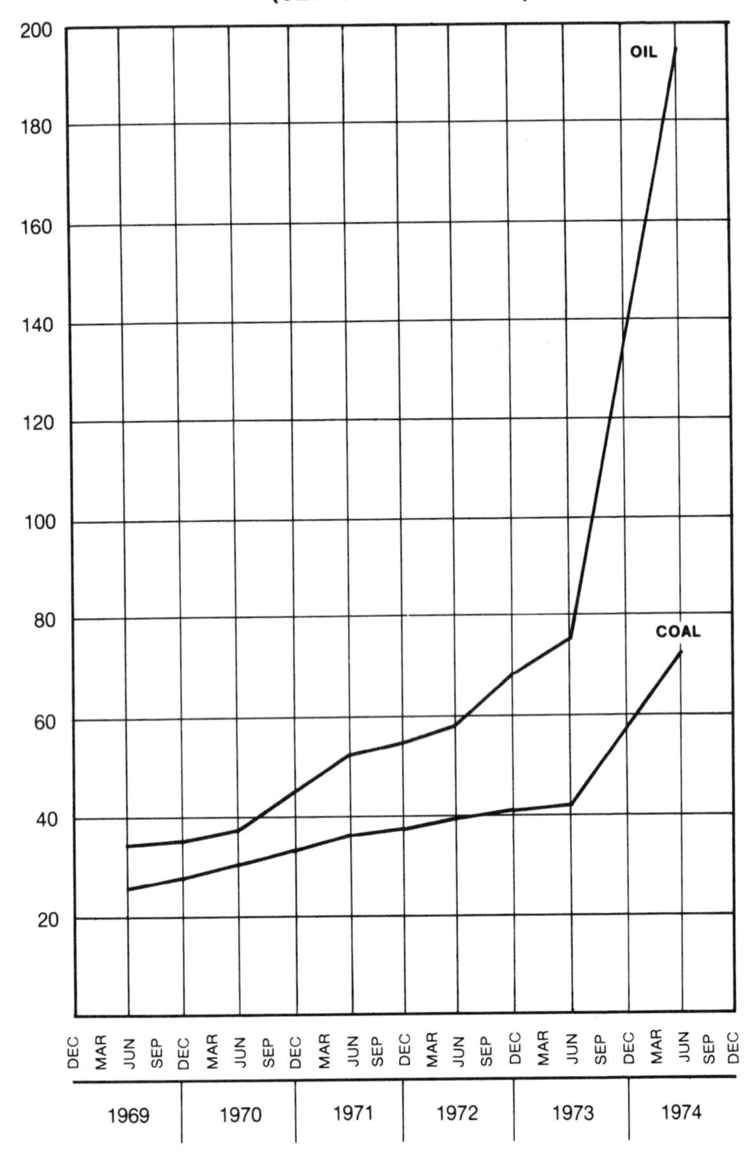

Source: National Coal Association

EXHIBIT 4

CONCENTRATION IN ENERGY PRODUCTION, 1974
(BTU Basis)

	Share of Total Energy Production (%)
Largest firm	6.8%
Top 4 firms	19 %
Top 8 firms	31 %

EXHIBIT 5

FORECAST OF ENERGY INVESTMENT

I. The Federal Energy Administration, for its Project Independence Blueprint, commissioned Arthur D. Little, Inc., to develop the investment outlays necessary to meet the energy demand conditions set out in detail below. Essentially, these capital investment figures assume a moderate growth in energy demand (A.D. Little's "low demand" case) and a stable level of oil and gas imports taken together through 1985.

II. **Demand Assumptions:**

Growth Rates in Energy Demand by Primary Source

	Historic 1960-73	Projected 1973-90
Oil*	4.4	1.0
Gas**	4.7	1.1
Coal	1.9	3.8
Nuclear	—	22.4
Hydro & Other	4.5	4.2
Total Energy	4.1	3.2

*Includes oil from shale and coal liquefaction.
**Includes gas from oil and coal.

Total Energy Demand by Primary Source

	1973	1980 Quad. Btu's	1990
Oil	35	37	41
Gas	23	25	28
Coal	13	18	25
Nuclear	1	6	28
Hydro & Other	3	4	6
Total Energy	75	90	128

Shares of Energy Market

	1973	1980	1990
Oil	46%	41%	32%
Gas	31%	28%	22%
Coal	18%	20%	19%
Nuclear	1%	7%	22%
Hydro & Other	4%	4%	5%
Total Energy	100%	100%	100%

EXHIBIT 5 (Cont'd)

Share of Imports

	1974	**1980**	**1985**
Oil (MM Bbls./day)			
Domestic Supply*	11.0	13.8	16.2
Imports	6.1	4.6	4.4
% Imports	36%	25%	21%
Gas (TCF/year)			
Domestic Prod.**	21.7	22.2	22.7
Imports	1.0	2.3	3.5
% Imports	5%	9%	13%

*Includes oil from shale and coal liquefaction.
**Includes gas from oil and coal.

III. **Energy Investment:**

Energy Investment Requirements 1974-1990

	Billion 1974 $
Electrical	490-569
Nuclear Fuel	19-30
Coal*	40-52
Oil and Gas	334-352
Solar	8
Geothermal	8
Municipal Waste Conversion	8
Total	906-1026

*Includes transportation and liquefaction and gasification facilities.

EXHIBIT 6

IV. **Coal Investment**

Coal Capacity Additions and Investment Costs

	1973	1980	1985	1990
		Million annual tons		
Annual Capacity	599	893	1019	1895

	1973-80	1981-85	1986-90
New Mining Capacity			
Underground Expansion	75	—	—
New Underground Mines	120	85	323
New Strip Mines	234	173	685
Total	429	258	1008
		Billions dollars	
Investment (1) (2)			
Underground Expansion @ $7/ton	0.5	—	—
New Underground Mines @ $20/ton	2.4	1.7	6.5
New Surface Mines @ $15/ton	3.5	2.6	10.3
Total	6.4	4.3	16.8

(1) Assumes upper range of possible annual costs per ton. Lower range costs are: underground expansion @ $5/annual ton, new underground mines @ $17.50/annual ton, and new strip mines @ $10.00/annual ton. Total investment cost to 1990 in this case would be $20.5 billion.

(2) Consol estimates of investment costs per annual ton of capacity are: New Eastern underground mine—$30/annual ton; Eastern surface mine—$30/annual ton; Western surface mine—$8/annual ton.

Average Mine Size and Number of New Mines*

	Mine Size: Current Avg.	Mine Size: New Mines	Additional Capacity Needed	Number of New Mines Needed
	Thousand tons per year			
Underground	150	2,000	528,000	264
Surface	100	5,000	1,092,000	218
Total	—	—	1,620,000	482

*Average size of new mines is based on Continental estimates.

17

Suggestions For Doubling Our Coal Production

STATEMENT OF DEWITT W. BUCHANAN, PRESIDENT, OLD BEN COAL COMPANY, BEFORE THE SUBCOMMITTEE ON ANTITURST AND MONOPOLY, COMMITTEE ON THE JUDICIARY, UNITED STATES SENATE — OCTOBER 21, 1975.

Summary: In order to meet the national goal of doubling coal production, several measures are suggested: allowing free entry into the coal business and erecting no barriers to capital investment; encouraging research and development, partly through federal funding; a viable program to lease federally held coal lands; and sensible environmental and safety legislation. One of the major consequences of the forced divestiture of coal operations from oil companies would be reduced investment resulting in lower future coal production.

DeWitt W. Buchanan speaks . . .

The Old Ben Coal Company, of which I am the President, is headquartered in Chicago, Illinois and operates mines in Illinois, Indiana, and Virginia. As the result of a merger in 1968 between Old Ben and The Standard Oil Company of Ohio ("Sohio"), Old Ben is now a division of Sohio. I am currently a Director of Sohio and have been since April of 1969.

I'm here today to talk about the coal business because that is what I know. I grew up in the coal business; I've always been in it, and I always will be. I understand that you have heard from a lot of different witnesses during these hearings, but as far as I could determine, you've not heard from anybody who's actually devoted his life to coal. I appreciate this opportunity to tell you something about the facts of life in this industry, past and present. I would also like to make some suggestions on what the government can do to help with the future development of coal in this country.

The Coal Industry

The United States is blessed with very extensive coal reserves as illustrated by the map of Exhibit I; some of these reserves are well located with respect to coal markets and others are not.

A recent publication of the U.S. Bureau of Mines entitled "Demonstrated Coal Reserve Base of the United States on January 1, 1974," states that this country has 434 billion tons of demonstrated coal reserves. Assuming conservatively that 50 percent of these reserves are economically recoverable with today's technology, this represents a future coal supply of over 200 billion tons. To put that into the proper perspective, the current rate of U.S. coal production is approximately 600 million tons per year. At that rate of production, the United States has approximately 350 years' supply of recoverable coal. Even if our annual production rises substantially, we have a lot of coal by any standard of measure. Stated another way, these coal reserves on a BTU basis are equivalent to almost one trillion barrels of crude oil. This far exceeds the estimated U.S. crude oil reserves of around 36 billion barrels.

The coal industry is made up of 4,000 operating companies of different sizes. In 1974, the largest company, Peabody Coal, produced 68 million tons of coal, or slightly more than 11 percent of the total production. During that year, 72 other companies had production in excess of one million tons. Additionally, 524 coal companies had production between 100,000 and one million tons. Over the last eight years, the sales of the 50 largest coal companies have fluctuated between 65 and 70 percent of the total industry's sales.

Over the years, a number of companies have entered the coal business and others have disappeared. There's nothing mysterious about this. As in any competitive industry, whether a specific company is able to survive and grow or not is a function of such factors as the quality of its personnel, how efficiently it invests its capital, and the quality and location of its coal reserves.

Historically, the health of the coal industry has been largely dependent upon factors which were beyond the control of the coal companies themselves. In 1940, the consumption of coal was heavily reliant upon use for home heat, railroad fuel, and the manufacture of coke. This can be seen in Exhibit II. In order to have an assured coal supply for coke production and other uses, many steel companies operated their own mines then and still do.

Following World War II, the rapid development of U.S. oil and gas production began to catch up with the coal industry. Due to the development and acceptance of the diesel locomotive, the percent of domestic coal consumption going to the railroad industry declined from approximately 22 percent in 1945 to less than 4 percent in 1955. (See Exhibit III.) By 1960, the railroad market for coal had essentially disappeared. Today, coal is not used in the transportation market at all.

Since the price of natural gas has been held below its true fuel value by price controls, and since petroleum products have been readily available, a major portion of the home heating market has shifted from coal to these cleaner and more convenient fuels. In 1950, 84 million tons of coal were used to heat homes. By 1974, home heating consumption of coal had declined by 90 percent to only 9 million tons.

Even though the U.S. demand for energy was growing rapidly during the 1950's and 1960's, the portion of that demand supplied by the coal industry was declining. Based on U.S. Bureau of Mines data, coal represented around 35 percent of the nation's total energy supply in 1950. By 1970, coal's contribution to the U.S. energy supply had declined to slightly less than 19 percent. The coal demand by the steel industry and a portion of the developing demand of the electrical utility business was all that kept the coal industry going.

The supply of coal to the utility industry has increased from 49 million tons, or 11 percent of total coal consumption in 1940, to a level of 388 million tons in 1974. The latter figure represents 70 percent of the total U.S. coal consumption in 1974. This apparently rapid growth in the demand for coal by the utility industry is very deceptive because the utility industry itself has been growing at a much faster pace.

Coal represented over 75 percent of the total fuel consumed in electricity production during the 1940's. This share of the utility market has been eroded to about 53 percent in 1974. Several factors have contributed to coal being squeezed in this market. Foremost among them has been the availability of cheap natural gas, and what I call dump priced imported residual fuel. When the import restrictions on very low priced residual fuel were lifted in 1966, large numbers of formerly coal-fired utility plants along the Atlantic Coast converted to residual fuel. Additionally, during the 1960's many utilities were reluctant to construct coal-fired facilities because nuclear power was being viewed as the fuel of the future in electricity generation.

Prior to the last two years, the wide availability of cheap natural gas and low cost imported residual fuel have enabled the utilities to more or less dictate coal contract terms. The Tennessee Valley Authority, for example, being the largest coal purchaser in the country, had the economic power to secure preferential pricing.

The continued regulation of natural gas prices at unrealistically low levels has caused an increase in demand for natural gas and has discouraged the development of additional supplies. Therefore, in the last few years, the availability of natural gas has fallen short of demand. Additionally, the cost of petroleum products has increased dramatically due to the pricing actions of the OPEC nations. These factors, coupled with a heightened concern over our growing reliance on imported oil, have brought the coal industry's potential back into focus. Coal is being called upon to play a major role in increasing the nation's energy self-reliance. A widely accepted goal is to at least double coal production by 1985.

The coal industry is more than willing to strive for that goal, but as in the past, the attainment of that goal is largely dependent upon forces outside of the industry's control.

As I pointed out in my discussion of the coal industry's history, both the transportation market and the home heating market have essentially been closed to coal. The utility industry represents the only portion of the energy market that can greatly increase its use of coal. This is in terms of both new facilities and the conversion of existing facilities which currently consume oil or gas. The extent to which coal can further enter this market is dependent upon the utility industry's capability to finance these expansions and conversions.

Additional external factors which have come about recently are also tending to inhibit the further development of the coal industry. These include the National Environmental Policy Act and the Federal Coal Mine Health and Safety Act. Please don't misunderstand. I am not against the good intentions for which these laws were designed. What I am against is the arbitrary manner in which they are being interpreted and enforced. They were well conceived measures, but they are currently being over-used and abused, as I will discuss more fully later.

From these brief remarks, I hope you will understand that the coal industry has been characterized for many years by three things: an abundance of reserves, the involvement of many companies, and the impact of what I have called external factors.

The History of Old Ben

Having mentioned some of the significant characteristics of the coal industry, let me turn to Old Ben itself and its relationship to Sohio.

Old Ben has been in the coal business for a long time. We are currently in our one-hundredth year of business. I'm very proud of that! Even though Old Ben was a public corporation prior to the merger with Sohio, it was family controlled. This company was founded by my grandfather. My father spent his entire career at Old Ben. I started with the company in 1940, and since then I have been involved in every single part of this business.

Over these one hundred years, we have developed a very good business reputation. This is important not only to our ability to market coal, but also in attracting top quality employees. One of the things I insisted on in the merger with Sohio was that Old Ben maintain its own identity. It has, and almost all of my key employees have come out of the coal side of the business.

At the present time, Old Ben operates three deep shaft mines in Illinois, one underground mine in Virginia, and two large surface mine complexes in Indiana. We are just bringing on stream a new large Indiana surface mine. In addition, we have commenced shaft sinking of a two-mine deep complex in southern Illinois.

Incidentally, our current plans for the new Illinois deep mines involve a capital expenditure of $80 million, exclusive of the coal lands cost. This is about $18 per annual ton of capacity. Twelve years ago we brought in a new coal mine in Illinois that entailed a capital expenditure of $12 million, also exclusive of the coal lands cost. That amounted to $4.60 per ton of annual production. As you can see, capital costs for coal mine openings have expanded fourfold over the 1965 to 1975 period.

Old Ben currently employs about 2,400 people at our mining operations. The expansions I have just mentioned will add another 1,450 in the next five years.

During the mid 1960's, we were looking for an opportunity for diversification in both Old Ben's geographical area of operation and its type of coal product. We were also looking for some step which would enable us to grow more rapidly in the coal industry. When Sohio approached us, therefore, we were interested. We felt that Old Ben would benefit from a merger with a relatively large, diversified and financially strong company like Sohio. In addition, I had a personal interest in Sohio's research and engineering, particularly the research it had done in the 1950's with respect to converting coal into synthetic fuels.

As far as Sohio's viewpoint is concerned, I understand that in the mid 1960's, their projections of energy supply and demand for the United States indicated that there would be a shortage of *total* energy during the decade of the 1970's. Their forecasts indicated that the production of crude oil in the United States would peak out some time between 1972 and 1974.

Sohio had long been a crude oil deficient refining and marketing company. At that time they were refining approximately 160,000 barrels of crude oil per day and their domestic crude oil production was in the range of only 30,000 barrels per day. Their previous efforts to improve this crude-deficit position had not been very successful.

Given this situation, Sohio's management was quite concerned about any opportunity for continued growth possibilities in the future. Therefore, in planning for the future they decided that if there was going to be a severe shortage of fuels, then other fuels should be good business for them to be in. This was particularly true of the one fuel which the United States had tremendous reserves, namely, coal.

Sohio felt that the only way they could enter the coal business would be to acquire a well-managed, medium sized coal company which had sufficient reserves for expansion and was reasonably profitable. The main guideline was that the company had to be technically competent and well managed.

If I have a capital project which appears to be profitable, I get the funds I need for that project. Old Ben's coal production has not been restrained by my friends at Sohio. In fact, I constantly get strong encouragement and support to produce as much coal as possible, because Old Ben's recent profits have added important support to Sohio's major financing efforts to develop Prudhoe Bay and construct the Trans-Alaska Pipeline.

Old Ben's coal prices have *not* been manipulated to support Sohio's oil prices. I have as much control over my prices as the competitive coal market will allow. In terms of operating authority, I can make any coal sale up to $5 million without even notifying Mr. Whitehouse. I don't need higher approval for any coal sale unelss the total revenue resulting from that sale exceeds $10 million. As you can see, I have a lot of latitude in this area.

I've touched on a number of the real plus factors in Old Ben's association with Sohio. In doing so, I guess I am disagreeing with some of your concerns.

Competition — Coal Versus Petroleum

I understand that one of the major things that worries this Subcommittee is the potential decrease in competition which might result from individual companies producing more than one form of energy. I personally feel that this concern is unwarranted, particularly in regard to petroleum products and coal.

I suggest that you consider the various markets served by petroleum products and coal, as well as the normal method of sale. These factors will confirm that coal is in extremely limited competition with petroleum products, regardless of production ownership.

Take a look at the transportation market. It's not surprising that this has become the exclusive domain of various petroleum products. In fact, about 56 percent of all refined petroleum products are consumed in highway, air, water, or rail transportation. Coal had been a significant factor in the railroad portion of the transportation market, but this market for coal had disappeared by 1960.

Approximately 18 percent of petroleum products end up in the home heating market. By 1974, less than 2 percent of the bituminous coal consumed in this country was used for home heating, and this percentage was continuing to decline rapidly. A substantial number of the new houses which have been constructed over the years have heating systems based on natural gas or electricity. These installations are designed such that it would be impossible to convert them into coal. With the convenience and cleanliness inherent in the use of petroleum products and natural gas for home heating, I find it difficult to believe that the few home owners who could convert their furnaces back to coal would do so as long as they can get natural gas or heating oil.

Both petroleum and coal can be converted into coke. This application represents 16 percent of U.S. coal consumption as opposed to less than 2 percent of refined petroleum products. However, these two types of coke serve entirely different markets. Coke produced from coal is used exclusively in the production of ferro-metals. Generally, petroleum coke is converted into high purity electrodes that are used for, among other things, the refinement of aluminum. Consequently, there is no overlapping market for these types of coke.

Eight percent of refined petroleum products goes into either petro-chemicals or asphalt. Coal does not have the physical characteristics needed for asphalt and the amount of coal consumed in chemical production is insignificant. Again, there is no overlap of markets in oil and coal used for these purposes.

A minor amount of refined petroleum products, namely 3 percent, is consumed as a boiler fuel in the industrial market. While industrial boiler fuel consumption of coal had been a very significant market until the mid 1960's, it has been dwindling since then. Due to the relatively high cost of air emission control devices necessary to meet environmental regulations, many industrial firms, particularly the smaller ones, have found the continued use of coal to be uneconomical. By 1974, industrial use of coal had declined to less than 12 percent of total coal consumption compared to the mid 1960's level of around 25 percent.

The only significant area of consumption served by both coal and petroleum products, specifically residual fuel, is the electrical utility market. By 1974, over 70 percent of the bituminous coal consumed in this country went to electrical utilities. Residual fuel sales represented 16 percent of U.S. petroleum product demand in 1974 and over 60 percent of this residual fuel was not produced here but was imported from overseas. That portion of residual fuel sales and the small amount of distillate fuel sales going to the electrical utilities in total represented 9 percent of refined petroleum product demand in 1974.

Even in the utility market, competition between coal and residual fuel is limited by geography, technology, and method of sale. The residual fuel used by utilities is essentially limited to the New England states, the Mid-Atlantic states, and the West Coast. This had been the result of a growing reliance over the years prior to the Arab embargo upon low cost imports of residual fuel.

Utility plants are very costly to construct. Therefore, the furnaces are typically designed to burn one type of fuel. In most cases, once a plant is constructed, it no longer acts as a market for fuel types other than that for which it was designed, unless extensive capital modifications are undertaken.

Since residual fuel is the product of a manufacturing step, it can be produced from a variety of refineries and still meet the same specification. Therefore, those utility plants which burn residual fuel can enter into short or intermediate term contracts, with some assurance that when the contract terminates, suitable residuals can be obtained from another refiner, if necessary.

This is not the case with coal. Each area of coal reserves has its own physical characteristics. Coal produced from one mine, while being perfectly suitable for use at one utility plant, may be completely unacceptable at another. Hence, the majority of coal sold to utilities moves under long term contracts which might last for the life of the mine. One of the objectives of the utility plants in doing this is to avoid costly equipment modifications which could be required every time a short or intermediate term fuel contract expires.

Typically, a utility planning to construct a coal-fired plant will issue specifications on the volume and type of coal required and the plant location. In the case of Old Ben, if we have reserves of the type required within a reasonable geographic distance from the consuming plant, we would attempt to negotiate a supply contract. Our success in winning this contract would depend upon our ability to supply an acceptable specification product at a lower cost than our competition. This is a direct function of how efficiently we are able to mine coal. If we do enter a contract, then we would construct a mine and essentially dedicate its production to that particular customer.

I would estimate that approximately 75 percent of the coal sold to electrical utilities is under long term contract. That coal which is not sold under long term contract moves in what is generally known as the spot market. Here competition among coal companies is very intense. Both the supply and demand for coal in the spot market are dependent upon general economic conditions. During recessions such as the current one, the amount of coal moving into the export market declines due to slackened demand. Therefore, more uncommitted or spot coal becomes available. However, demand for spot coal also is down. Recently, my sales department has been crying on my shoulder every day because they are finding it very difficult to sell uncommitted coal.

When economic conditions improve, export volumes increase and less coal is available for the spot market, just at the same time as demand for that coal increases. That is why spot coal prices fluctuate so much, compared to contract prices which remain relatively stable over long periods of time. This can be seen in Exhibit IV.

S. 489

The bill that this Subcommittee has under consideration, S. 489, would prevent companies within the petroleum industry from engaging in the development of other forms of energy. In doing so, it would force petroleum companies to divest themselves of their coal interests. This would result in a major structural change for some coal companies. Structural change of a company or industry without proof of a need for change or a consideration of what will really result doesn't make a great deal of sense to me.

I disagree with the thrust of S. 489 because it appears to be directed toward imaginary problems related to "potential" limits to competition between energy forms. I believe that these problems just do not exist. As I have indicated in my discussion of the relationship between Old Ben and Sohio, Old Ben's coal production has not been held down by Sohio to stimulate demand for petroleum products. *In fact, our coal production has increased substantially.* Nor has Sohio attempted to manipulate Old Ben's coal prices.

I feel that enactment of S. 489 would not change the limited nature of competition between coal and petroleum products. What it would do is disrupt the widely recognized efforts to increase all forms of domestic energy production, including coal, because it would eliminate the positive aspects of petroleum company ownership of coal operations.

Look at what has taken place at Old Ben since its merger with Sohio. Since 1969, Old Ben as part of Sohio, has committed $155 million for development, coal reserves and surface land, and physical plant and equipment. During a similar time period immediately prior to the merger in 1968, Old Ben had made capital expenditures equal to about $30 million.

During the five-year period from 1964 through 1968, Old Ben's annual coal production had averaged a little over 8 million tons. Since 1969, Old Ben's coal production has averaged over 11 million tons per year. The two new mining complexes we have under development will produce 7.5 millions tons annually.

The real question is whether coal production would expand faster if oil companies were ordered to divest their coal interests. I think just the opposite would be the case. There would be expansion, but very likely it would be at a much slower rate. Let me speculate on what might happen if Old Ben were divested. With construction costs going up so rapidly, it is obvious that financing has become a critical factor in expanding coal production. Coal companies can normally offer lenders the security of known coal reserves and of a market for the coal by means of long term contracts. However, developing a mine is a risky business. Therefore, lenders normally require the additional assurance that the mine will, in fact, be completed and kept in operation until the loan is repaid. If a mine shaft fails, for example, another one must be sunk. In our case, Sohio provides this guarantee to the lenders.

If we become a separate company, we could provide such guarantees to the limit of our financial ability. However, this would slow down our rate of expansion until we became larger as time passed. We would probably have to pay for mines by selling stock in the company, and this is an expensive way of financing such a development. In any case, in the near term, we could not expand at a pace which would allow us to contribute our share to the goal of doubling coal production in the next decade.

Consequently, I feel that enactment of the legislation which you are considering would simply disrupt a well organized effort to increase coal production. I do not feel that restructure of the industry would accomplish anything in a positive sense. S. 489 simply does not address itself to the real problem of the coal industry. I would like to comment on several things that do cause a problem for this business.

Our current environmental regulations are designed to limit the amount of pollution entering our air and water. Obviously, there are a number of ways that this can be accomplished. Unfortunately, in regard to air emissions, the regulators have insisted upon artificially imposed restrictions which must be met at all times. In doing so, they have ignored the limited availability of technology to meet these restrictions, the costs involved, or whether intermittent controls which operate only when required by climatic conditions would be more appropriate. These artificial restrictions must be loosened in the short run if the coal industry is to provide a greater portion of this nation's energy supply. Congress must restore a proper balance between the needs of the economy and the needs of the environment.

Similarly, the Coal Mine Health and Safety Act (CMHSA) has had a very adverse impact on coal production without a commensurate increase in safety. Since this law was passed in 1969, the productivity of underground mining has been reduced by as much as 45 percent. I wish that the coal industry and the people administering CMHSA could work cooperatively to improve safety without substantially decreasing productivity. However, the manner in which CMHSA is being interpreted and applied is inhibiting progress towards achieving the intent of the Act.

One of the major steps necessary to increase coal production in this country is to engage in applied research on mining techniques which will improve coal productivity with increased safety. Yet the rigid interpretation of CMHSA that is being applied by the Interior Department has all but precluded such applied research. The coal industry-supported research organization, Bituminous Coal Research, is currently being reorganized to provide leadership in an effort to correct this situation.

As I said earlier, I'm for measures which improve mine safety, but I'm against the current interpretation of CMHSA which is holding down the production and use of coal. Now Congress has had a say in enacting this law, and I believe it can have a say in how it should be interpreted.

If you are interested in how to encourage greater coal production, I have the following suggestions to offer:

1. The coal industry must be allowed to attract the capital needed to expand coal production. Estimates of the amount of investment required to double coal production between now and 1985 have ranged from $15 billion to $20 billion. Generating that much capital is going to be quite a challenge to the industry. I don't feel that the coal industry should be assisted in attracting this capital through either federally guaranteed prices

or federally guaranteed loans. At the same time, capital formation should not be hampered by barriers designed to keep specific segments of the economy out of the coal business. The more companies that enter this business, the more capital there will be. Coal production should increase accordingly.

2. Research and development in the coal industry should be strongly encouraged and Federal funding should continue. I realize that a great deal of research is being done on the liquefaction and gasification of coal. Economically feasible synthetic fuel processes are important but will be of little value if the coal they need for feedstock is unavailable. The critical research required is on how to increase productivity in existing and future coal mines. New underground mining methods have to be developed to supply the feedstock for these synthetic processes.

3. A viable Federal coal leasing program must be established which will allow a business-like development of western coal. In order to allow the smaller coal operator to participate in this development, bonus bidding for these leases should be replaced by royalty bidding. I personally feel that the Interior Department is headed in the wrong direction in some of its proposals with respect to the development of these coal lands. I am opposed to a program such as the IMARS program proposed by Interior which dictates what shall be leased and by whom.

4. A realistic means of complying with NEPA must be established so that the increased production and use of coal can proceed without undue delay or restraint. The amount of time involved preparing environmental impact statements and getting them approved, usually through judicial review, must be shortened. The current procedure delays both the opening of new mines and the siting of coal-fired plants.

5. The Clean Air Act amendments which have been proposed by the President should be enacted.

6. No Federal reclamation legislation should be enacted which is so unreasonably stringent that it precludes the surface mining of western lands. Reclamation has become a well established practice by the responsible elements of the coal industry. Additionally, the individual states have developed reclamation policies designed for their own lands. A Federal program which attempts to deal with wide variations in topography in a uniform manner would not only be redundant but would severely retard coal production.

7. Methods to improve and expand coal transportation facilities should be taken into consideration. Significant expansions of coal production will require a healthier transportation system than the one that exists today.

8. Oversight hearings should be conducted on the Coal Mine Health and Safety Act. In particular, the provisions pertaining to the assessment of fines and judicial review need attention. The time and effort presently involved by both Government and management on account of these provisions of the Act could better be devoted to actual efforts to achieve safety and productivity.

A Final Word

I would like to make one final point here today. I understand that various witnesses who have appeared before this Subcommittee have implied, without adequate factual justification, that there is little competition within the petroleum industry due to joint ownership of facilities, alleged interlocking directorates and the like. I came into this industry as a complete outsider seven years ago. I've been a member of Sohio's Board of Directors for the past six years, and I'm happy to say if Sohio's operations are typical, these allegations are simply not true. These unsubstantiated assertions do a disservice to the companies working to meet the energy needs of the nation.

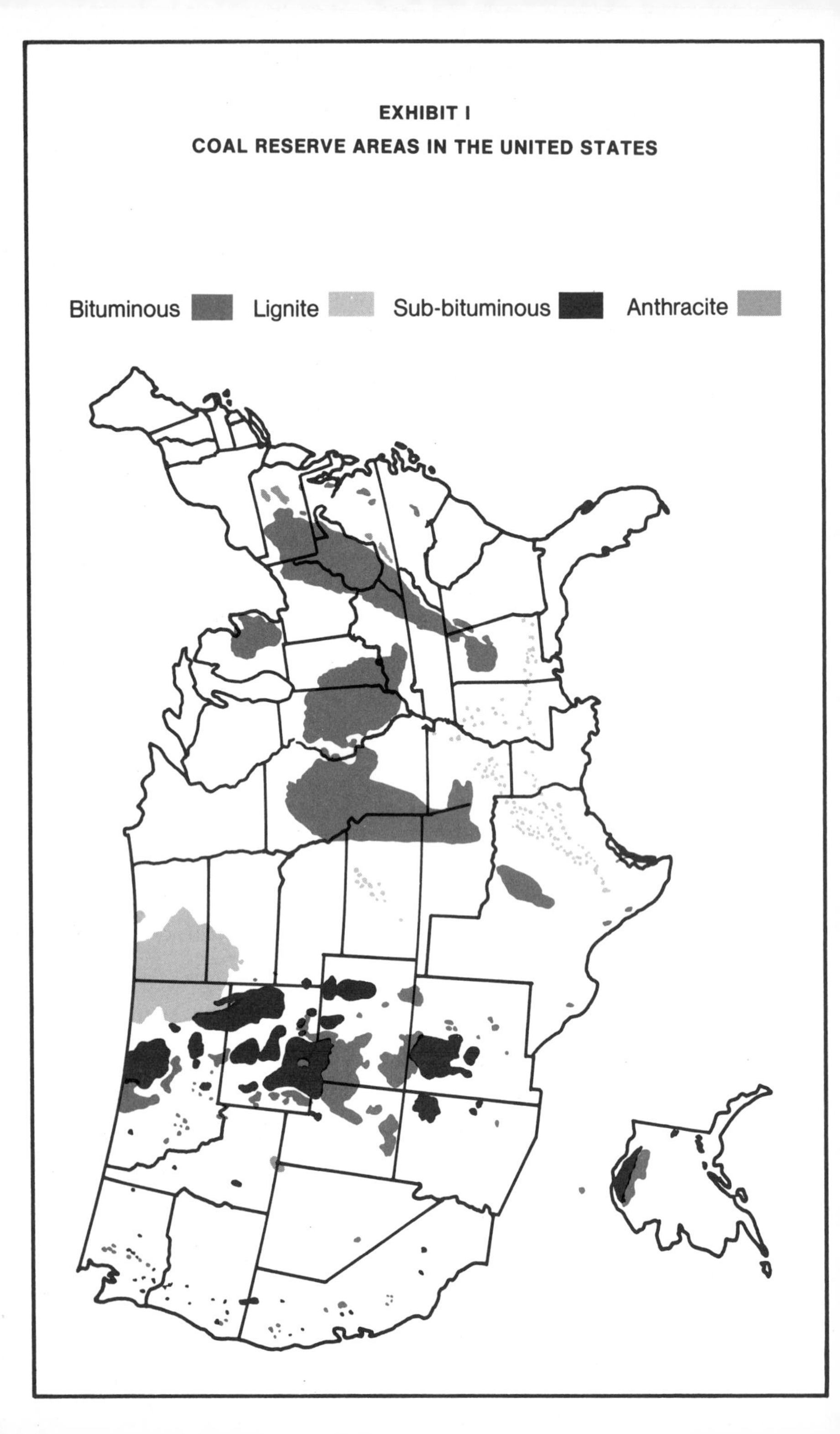

EXHIBIT I

COAL RESERVE AREAS IN THE UNITED STATES

EXHIBIT 2

United States Consumption of Bituminous Coal for Selected Years

(Millions of Tons)

Year	Railroads	Electric Utilities	Coking Coal	Other Industrial	Retail Deliveries	Total
1940	85	49	81	131	85	431
1945	125	72	95	149	119	560
1950	61	88	104	117	84	454
1955	15	141	107	107	53	423
1960	2	174	81	93	30	380
1965	—	243	95	102	19	459
1970	—	319	96	89	12	516
1974 (1)	—	388	90	64	9	551

(1) 1974 Figures are preliminary

Source: *Bituminous Coal Data,* 1974 Edition, published by the National Coal Association, page 83.

EXHIBIT 3

United States Consumption of Bituminous Coal for Selected Years

(Percent)

Year	Railroads	Electric Utilities	Coking Coal	Other Industrial	Retail Deliveries	Total
1940	19.7	11.4	18.8	30.4	19.7	100.0
1945	22.3	12.9	17.0	26.6	21.2	100.0
1950	13.4	19.4	22.9	25.8	18.5	100.0
1955	3.5	33.4	25.3	25.3	12.5	100.0
1960	0.5	45.8	21.3	24.5	7.9	100.0
1965	—	52.9	20.7	22.3	4.1	100.0
1970	—	61.8	18.6	17.3	2.3	100.0
1974 (1)	—	70.4	16.4	11.6	1.6	100.0

(1) 1974 Figures are preliminary

Source: *Bituminous Coal Data,* 1974 Edition, published by the National Coal Association, page 83.

EXHIBIT 4

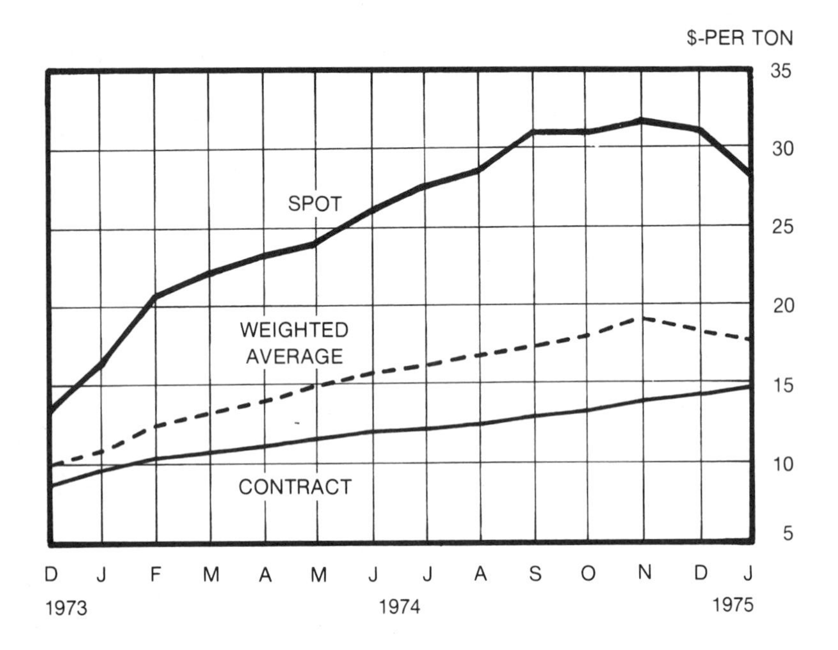

Source: Federal Power Commission

18

The Know-How To Develop Oil Shale

TESTIMONY OF JOHN E. KASCH, VICE PRESIDENT, STANDARD OIL COMPANY (INDIANA), BEFORE THE SUBCOMMITTEE ON ANTITRUST AND MONOPOLY, COMMITTEE ON THE JUDICIARY, UNITED STATES SENATE — OCTOBER 22, 1975.

Summary: The future course of development of the United States' huge shale oil resources is characterized by uncertainty, the need for enormous capital investments, and further technical and engineering progress. There is no evidence that oil industry participation in oil shale development has had anything but a positive impact on competition or the public interest. Further, forced divestiture of oil companies' oil shale interests would probably bring to a halt oil shale projects currently in progress.

John E. Kasch speaks . . .

Standard of Indiana, along with the Gulf Oil Corporation, is a coholder of an oil shale lease issued in early 1974 under the Department of Interior Prototype Federal Oil Shale Leasing Program pursuant to competitive bidding. One of my overall responsibilities at Standard of Indiana is the development of this new source of domestic petroleum—shale oil. It is a tremendous challenge and one fraught with economic, environmental and technical problems.

From the technical viewpoint, the technology for the recovery and upgrading of raw oil shale into high-quality crude petroleum, while quite similar to conventional petroleum refining technology, has never been demonstrated on the scale necessary for commercial production. The equipment is certain to be complex and very costly, particularly in the early stages of development. Environmental groups, rightfully, are concerned about the impact of an oil shale industry on the Western United States. To accommodate that concern, and under the terms of the lease, we are carrying out intensive base-line environmental studies and we believe that this development can be accomplished in an environmentally acceptable manner.

The Federal, state and local governments are interested and concerned in many aspects of oil shale development, in fact, over 200 agencies of these government bodies have an expressed interest of some type. Satisfying such a large group will be difficult and it is not possible to predict the governmental regulations which may be imposed on our operation.

Finally, from an economic viewpoint, oil shale development is highly capital intensive and the price obtainable for the product is uncertain, thus the economic risks are very high.

But we remain optimistic that we will be successful in the development of this new petroleum source for energy consumers of the United States. The stakes are high, but left room to exercise our ingenuity and resourcefulness, we will get the job done.

Today, I plan to restrict my discussion to oil shale. Others will discuss coal, geothermal, nuclear energy and the other fuels cited in S. 489. With our particular collection of talented people, technology, special resources and know how, we are devoting our efforts to oil shale at this time because we believe we can compete successfully in this development. Indeed, this is the basic consideration for our getting into any venture—can we compete successfully.

Our interest in oil shale also stems from the belief that the addition of this vast petroleum resource to our domestic reserves can materially assist in solving the nation's energy problems.

Oil shale in the United States is a huge untapped natural resource. Its primary value is that an oil virtually equivalent to crude petroleum can be obtained by a complex process of mining, retorting and upgrading. The production cost is high and shale oil could not compete in the past with abundant domestic and imported crude oils. Domestic crude production is declining and imported crude oil is very costly and not always readily available, consequently, there are both economic and national security incentives to develop a U.S. oil shale industry.

The largest known deposit of oil shale in the world is the Green River Basin of Colorado, Utah and Wyoming. About 600 billion barrels of oil are potentially recoverable by processing techniques under development. By contrast, this is approximately twice the proven reserves of the Middle East and more than a 100-year supply of oil for the U.S., at present rates of consumption. Crude oil reserves in the U.S. totaling less than 35 billion barrels last December 31, are equivalent to only about nine years' supply.

Oil shale development in the United States has been in progress since the 1850's. The history of attempts to establish commercial oil shale production has been that of a series of boom and bust cycles. A number of small pilot plants established prior to 1920 became noncompetitive because of lower priced East Texas crude oil and discontinued production. As a result of the Synthetic Fuels Act, the Bureau of Mines created and operated a small facility at Anvil Points, Colorado from 1944 to 1955. In the years following, the Sinclair Oil Company, the Union Oil Company, TOSCO, Mobil and Equity Oil carried out research and development efforts on the retorting of oil shale. In 1964, a group of six petroleum companies (Mobil, Standard of Indiana, Continental, Phillips, Humble and Sinclair) leased the Anvil Points facilities and carried out development work on the Bureau of Mines Gas Combustion Retort.

Later, the Colony Development Operation of Cleveland Cliffs, Sohio, TOSCO and Atlantic Richfield successfully demonstrated the TOSCO II process at 1000 tons per day. Current research and development efforts include the 17 company Paraho Project and the 10 company WESTCO group; we are participants in both projects. The Paraho development is being carried out at Anvil Points in a continuation of the development of the Gas Combustion Retort process. The WESTCO project is oriented towards in-situ retorting.

The point illustrated by this history of oil shale development is that the specialized talent necessary came from the free entry of firms that drew heavily upon their previous experience. The Project Independence Report of November, 1974, lists no less than 71 "firms... with identified interests in oil shale development in Colorado" alone. A number of these are oil companies which have made the major investments in research and development. However, a large number of the companies on the list are from other industries. This indicates that companies from many different industries may be expected to participate in oil shale development when the pioneer stage is past. A copy of that list is attached for your information.

The oil shale lands are largely owned by the U.S. Government. In addition, many individuals and companies own or hold interests on oil shale property outside the Federal Leasing program. To expedite development, the Department of the Interior offered six new leases of approximately eight square miles each in early 1974. Four leases were awarded in open competition with Standard of Indiana and Gulf Oil Corporation successfully bidding just over $210 million for the first lease in Colorado known as Tract C-a. You can appreciate that with a commitment of that magnitude, we are anxious to proceed to production as quickly as it is economically feasible. If that were not enough, the lease contracts provide benefits and penalties that encourage development of prototype units.

The Rio Blanco Oil Shale Project is a joint undertaking of Gulf and Standard and is headquartered in Denver, Colorado with an office in Rangely, Colorado; tract C-a is near Rangely. During the past year Rio Blanco has had numerous contractors exploring the area to establish baseline data on vegetation, animal life, water and air quality to assist in protecting the environment. In addition, numerous core holes have been drilled to define oil shale quality and location and to seek on-site water.

Oil shale, being a rock-like material, requires mining as the initial step in the oil recovery process. Open pit mining is feasible on tract C-a because the overburden at some points is only 80 feet thick. At other locations, the overburden reaches 800 feet. A pit would be created similar to that produced by iron ore mining in Minnesota or copper ore mining in Utah. The pit would be filled in with retorted shale towards the end of the project to restore and improve the original contours of the land. Our present plans project recovery of 60 percent of the oil in shale richer than 15 gallons per ton during a 57-year open-pit project.

Shale oil production by in-situ retorting has been under study for many years. Currently, in-situ techniques are being developed by the Occidental Petroleum Corporation, the Lawrence Livermore Laboratories and by the Western Oil Shale Corporation in a multicompany project. In these experiments, a portion of the shale is mined to create a void space and the rest rubblized by explosives. The resulting underground column of shale containing voids makes below-surface combustion possible so that hot gases heat the shale to retorting temperatures. The oil is collected at the bottom of the rubble, and pumped to the surface in a manner similar to a conventional oil well. The mined shale which is produced will be retorted above ground. Processing of shale in-situ may provide a method of shale oil recovery that consumes less capital, manpower, energy and water. Work on this technique is continuing.

Shale oil is essentially equivalent to crude petroleum, and the processing technology is similar and has been demonstrated both in the laboratory and in a recent refinery test sponsored by the Energy Research and Development Administration in Colorado. The processes used include visbreaking, hydrotreating, hydrocracking, catalytic cracking and coking that are very familiar and widely used by the petroleum refining industry. Evaluation of the products from this test were made primarily by the U.S. Navy and found to be satisfactory.

An oil refiner switching to oil shale is analogous to a textile mill switching from cotton to rayon to polyester. These are all fibers that are candidates for producing cloth. Crude oil from shale is readily acceptable by refineries that have in the past used conventionally produced crude oil.

Shale oil is not a new business entry or an extension of integration, but merely a change in the source of refinery feedstocks.

I would like to say something concerning the effect of oil shale development on the local communities involved. As pointed out in the Task Force Report on Oil Shale in Project Independence, there will be sizable influxes of people into areas which are now sparsely populated as oil shale development takes place. For our part, we are now actively working with the Bureau of Land Management, the State of Colorado, the County of Rio Blanco, and the nearby communities to handle the population influx into the region. Drawing on the experience of our partner, the Gulf Oil Company, with the new town of Reston, Virginia, we are assisting in the orderly development of the infrastructure of the area.

I want to stress at this point the tremendous costs in developing a prototype oil shale project. As of May of this year, our company and Gulf had already paid $84 million in installments in our $210 million bonus commitment with 37-1/2%, about $31 million, going to the State of Colorado. We have spent more than $12 million in environmental mining and processing studies, and we have committed ourselves to licensors and contractors for an additional $6 million. And yet we have hardly begun. We are in the stage of completing the ecological and environmental protection studies required by our lease. We are also preparing the final development plan for submission to the Government for approval in early 1976. Only upon receipt of that approval can we begin preparing the site and building facilities and developing the mines. We would hope to begin commercial oil production in early 1980 and reach full production in 1982. The full magnitude of investment for a 50,000 barrel per day plant will be in excess of $1 billion. We anticipate that it will cost over $100 million each year to operate at this level. We believe that our leasehold is capable of supporting facilities to produce at an ultimate rate of 300,000 barrels per day. Investment to achieve such production could exceed $4 billion.

We are not certain of the investment requirements for the three tracts leased under the prototype program to other companies, all of them in the oil industry. However, we would assume that both the investment required and the lead time before commercial production are similar to our own.

I would now like to address a few words to the probable impact of S. 489 and similar legislation upon oil shale development. S. 489 would forbid the entry of oil companies into the oil shale business and would require oil companies already engaged in oil shale activity to divest themselves of oil shale leases and presumably related facilities within three years after its enactment or suffer fines and/or imprisonment of its officers and directors.

Forced divestiture of leaseholds and other assets of oil companies currently engaged in oil shale development would, of course, severely penalize them. I have mentioned the millions of dollars expended to date by our company under the prototype program and, of course, we have incurred research expenses for many years. Of greater importance, however, divestiture would in all probability bring to a halt oil shale projects currently in progress. Whether new oil shale projects could be launched without the participation of those companies who have thus far shown themselves most interested in the field and whose technical abilities are the most complementary to oil shale development is doubtful. In any event, the risk of the loss of shale oil as a substitute for imported crude oil and its resultant effect both upon the American consumer and our national security are tremendous.

Further, the three states of Colorado, Utah and Wyoming would be deprived of benefits expected to flow from the development of oil shale. Planned investment benefits and increases in direct employment would be unlikely to occur. A broader tax base, new roads, communities and businesses to serve the proposed population centers would be less likely to materialize.

I must confess that I am unable to understand the reasons behind S. 489 and similar legislation. I know of no evidence that oil industry participation in shale oil development has or will have any adverse impact upon competition or the public interest. While it is true that members of the oil industry have been the leaders in investing in the development of oil shale technology and the preparation for commercial operation of oil shale reserves, there is nothing unusual or sinister about this. As I have stated, shale oil is simply a high grade feedstock for conventional petroleum refineries. The oil companies were among the first to question our nation's dependence on imported crude oil and to urge both the expansion of domestic oil and gas production and the development of alternate energy sources. The search for crude oil and natural gas is itself a risky business and the oil companies were deterred less than other industries by the equal or greater risks of oil shale production. In fact, it would have been somewhat surprising if the oil industry had ignored a major source of future feedstocks for the needs of the nation for petroleum products.

The fact that few companies outside the oil industry have aggressively pursued oil shale development does not mean that this will always be the case. Many companies other than oil companies hold interests in oil shale lands today. Nearly 500 of the 600 billion barrels of high quality oil shale reserves are on Federal lands. Thus far, less than 5 percent of these reserves have been leased under the Prototype Federal Oil Shale Leasing Program. If that program proves that oil shale can be developed economically while protecting the environment, it can be expected that other companies, unwilling at this time to accept the risks of developing this resource will be attracted into the field.

It should be noted that under existing Federal legislation, no more than one oil shale lease may be granted to any one person, association or corporation. While we consider this provision unduly restrictive, it does assure the availability of future leaseholds to large numbers of qualified applicants.

The entry of oil companies into oil shale development will not result in an anticompetitive situation. It is the natural result of technological strengths and raw materials requirements of the nation's energy consumers which our industry serves. Efforts of oil companies which have engaged in oil shale development are designed to increase domestic refinery feedstocks. If such efforts are successful, it will help not only to preserve competition among producers of such feedstocks, whether within or outside the oil industry, but also to contribute to our nation's independence from petroleum imports.

EXHIBIT I

Major Private Firms with Oil Shale Interests

Fuelco Resources Development, Gelco Company, Mitchell Energy and Development Corp., Petro-Lewis Corporation, Pittsburg & Midway Mining Company, Routt Mining Corporation, Seneca Coals Limited, Silengo Coal, Energy Coal, Colowyo Coal, Jenkins and Mathis Coal, Kirby Petroleum Company, Kimbark Exploration Company, Hanna Mining Company, Reliable Coal and Mining.

Chorney Oil Company, Northland Resources, Inc., Sundance Oil Company, Texaco, Inc., Mountain Fuel Company, Atlantic Richfield Company (La Sal Pipeline Co.), Tenneco, Union Carbide Corporation, Colorado Tungsten Corporation, CF&I Steel Corporation, Calvert Drlg. & Production, Development Services Corporation, Fuel Resources Development, Petroleum Information, Inc., Grassy Creek Oil Field.

Grassy Creek Coal Company, Pubco Petroleum Corporation, Mobil Oil, American Metals Climax Molybdenum Company, American Smelting & Refining Company, The Hanna Mining Company, CER Geonuclear, Four Mile Coal, Inc., Shell Oil Company, Chevron Oil Company, Union Oil Company, Standard Oil Company (Indiana), Getty Oil Company, Superior Oil Company.

Colony Development Corporation, The Oil Shale Corporation, Occidental Oil Company, Potrero Oil Company, Pro-Chemo Oil & Gas, Incorporated, Peabody Coal Company, Rocky Mountain Gas Company, Inc., Willard Pease Oil & Gas Company, Belco Petroleum Company, Colorado Interstate Gas, Virginia-Colorado Development Company, Kerogen Oil Company, D. A. Shale, Incorporated, International Nuclear, Exxon.

Bell Petroleum, Equity Oil, Cities Service, Tenneco, Continental, Sohio, Ashland Oil, Colorado-Ute Electric Association, Wolf Ridge Minerals Corporation, Public Service Company of Colorado, Paraho Oil Shale Demonstration Project (Anvil Points).

19

Drilling Expertise and Geothermal Energy

TESTIMONY OF CAREL OTTE, VICE PRESIDENT AND MANAGER OF THE GEOTHERMAL DIVISION OF UNION OIL COMPANY, BEFORE THE SUBCOMMITTEE ON ANTITRUST AND MONOPOLY, COMMITTEE ON THE JUDICIARY, UNITED STATES SENATE — OCTOBER 22, 1975.

Summary: Subsurface water heated by the natural heat of the earth brought to the surface becomes geothermal steam, which can be used to generate electric power. However, this energy production is possible in only a few places in the U.S. Geothermal energy production, now an infant industry fraught with technical and engineering problems, would require large investments. The technical, capital, and management resources of the petroleum industry are eminently suitable for furthering the development of geothermal energy. Conversely, prohibiting oil firms from engaging in this activity would inhibit or halt its growth.

Carel Otte speaks . . .

I have been actively engaged in geothermal work since 1962, and have personally participated in research and operating activities in most of the major geothermal areas of the nation. I am also active in scientific and industry affairs concerning geothermal work, and currently serve as Chairman of the Geothermal Advisory Group of the Energy Research and Development Administration. In addition, I am co-editor of the book *Geothermal Energy: Resources, Production, Stimulation*, published by the Stanford University Press in 1973.

As geothermal energy is not very well understood, perhaps a brief background comment may be in order. It is, in the broadest sense, the natural heat of the earth captured by means of subsurface fluids. The mass of molten rock or magma found below the earth's crust is usually too deep for its energy to be useful, but in a few areas it works itself close to the surface. Any underground water present also will be heated and will rise to the surface, sometimes causing such phenomena as hot springs, geysers or fumaroles.

We harness geothermal energy by drilling wells into potentially productive zones, much as we do for natural gas, and bringing the geothermal fluids to the surface. The steam itself cannot be transported over much more than a mile without losing pressure and temperature. Therefore, the most practical use for this energy is to turn electrical generators that are installed on the site of the geothermal deposit.

The drilling of geothermal wells is plagued with problems of hard, tough rocks, crooked holes, and formation fluid pressures which can cause lost circulation conditions. Naturally occurring gases, such as carbon dioxide, can lead to exceedingly corrosive conditions, which may affect the steel drill pipe and other equipment.

Nonetheless, much progress has been made in the drilling of geothermal wells. At The Geysers in Northern California, wells are drilled regularly to about 9,000 feet; in other areas to 7,000 and 8,000 feet. Yet much more needs to be done in all areas; improved metallurgy, better tools, better drill bits, new drilling fluids, new cement technology and whole new drilling procedures. As you can see, the oil companies already are well experienced in technical areas.

Our Nation's major geothermal development is The Geysers field. The development began in 1960 with a 12,500 kilowatt generating plant. In 1973 it became the largest geothermal development in the world, with a capacity of 400,000 kilowatts. Last spring another 100,000 kilowatts went into operation bringing the installed generating capacity to 500,000 kilowatts, sufficient to supply the electrical requirements of a typical city of 500,000 people.

Magma Power and Thermal Power (now a subsidiary of Natomas) have been involved in the field since 1957. In their first 10 years of effort, they produced steam enough to generate about 50,000 kilowatts of electricity. Getting this far was a slow and painful experience, stretching their technical and financial resources to the limit.

In 1965, based on a geological concept, adapted from our experiences in natural gas reservoir engineering, Union acquired a leasehold position in the vicinity of Magma-Thermal's lands. This concept was confirmed in

1966 by the drilling of our first discovery well. After this demonstration, which considerably enlarged the potential size of the field, Magma-Thermal recognized that their limited resources would not permit development of their holdings in a timely manner. To accomplish more rapid development of the field, Union and Magma-Thermal entered into a joint venture in 1967, with Union introducing its technological know-how, financial resources and management skills, and becoming the field operator. This means Union conducts the exploration, drills the wells, constructs the pipelines, produces the steam and delivers it to Pacific Gas and Electric Company, which then generates the electricity.

The same economic barriers that confront most infant industries or technologies are besetting geothermal operations. Drilling is a good example. Since we are drilling in harder rocks, with higher temperatures and often more corrosive fluids, it costs about 50 percent more to drill a geothermal well than a typical onshore oil or gas well of comparable depth.

Another problem is the very large capital investment required over a several year period before revenue flow can begin for a geothermal project. This is what limited the growth of Magma-Thermal in The Geysers in the early years. Utility companies usually need assurance of a 20- to 30-year life for a generating plant before regulatory agencies will permit construction. This, in turn, requires extensive drilling and well testing to prove the size and productivity of the reservoir. After the reservoir's capability is established, it can require up to four years to engineer and construct a generating plant.

The outlook for geothermal energy production has been studied extensively in recent months, especially by Federal agencies, including the Federal Energy Administration and the Energy Research and Development Administration.

The consensus seems to be that there is the geological opportunity for up to 20 million kilowatts of electrical generating capacity by 1985. This would be equal to five percent of current national electrical capacity.

To find and develop this capacity will require the drilling of at least 1,000 exploratory wells and 6,000 development wells, at a minimum average cost of $500,000 per well, or a total of $3.5 billion (in 1975 dollars) in drilling costs alone. Capital investment in hook-up facilities will add another $2 billion. Some 3,000 replacement wells will be required which, with the attendant capital investment, will bring the total investment requirement to some $10 billion.

These investment requirements are awesome sums of money, and I fail to see where they will come from if the oil industry is denied the opportunity to participate. Further, outside financial sources, even if available, will dry up if the technical expertise to spend these sums wisely and prudently is not available.

There is no doubt that geothermal development can be a significant factor in the effort to change our current energy consumption patterns. We burn too much valuable oil and natural gas in utility boilers. By backing oil and gas out of the boiler feedstream, we can achieve a double gain of reducing both imports and domestic oil and gas consumption.

Based on my years of experience in starting this new industry, I can only conclude that Senate Bill 489 will hurt rather than help the development of geothermal energy. For this job we need organizations that are venture-oriented, technically skilled, and are able to aggregate large sums of capital. These certainly must include oil companies. To do otherwise is to delay its needed development and to raise its ultimate cost.

Index